T0225064

Approximation of Euclidean Metric by Digital Distances

Jayanta Mukhopadhyay

Approximation of Euclidean Metric by Digital Distances

 Springer

Jayanta Mukhopadhyay
Department of Computer Science
and Engineering
Indian Institute of Technology Kharagpur
Kharagpur, West Bengal, India

ISBN 978-981-15-9900-2   ISBN 978-981-15-9901-9   (eBook)
https://doi.org/10.1007/978-981-15-9901-9

This Springer imprint is published by the registered company Springer Nature Singapore Pte Ltd.
The registered company address is: 152 Beach Road, #21-01/04 Gateway East, Singapore 189721,
Singapore

To my father Shree Dulal Krishna Mukherjee with deep gratitude, love, and respect.

Preface

In this monograph, different types of distance functions in an n-D integral space are discussed to consider their usefulness in approximating Euclidean metric. It discusses the properties of these distance functions and presents various approaches to error analysis in approximating Euclidean metrics. The main emphasis of this book is to present the mathematical treatises for performing error analysis of a digital metric with reference to the Euclidean metric in an integral coordinate space of arbitrary dimension. I hope that the monograph will be useful to researchers and postgraduate students in areas of digital geometry, pattern recognition, and image processing. The theory and results on the properties of different distance functions presented may have applications in various pattern recognition techniques. Analytical approaches discussed in the book would be useful in solving related problems in digital and distance geometry. As a prerequisite, the author expects that the readers have gone through first-level courses on vector algebra, coordinate geometry, and functional analysis.

There are six chapters in this book. In Chap. 1, the mathematical background of metrics, norms, distance functions, and spaces is presented with a brief discussion on the motivation behind the efforts in approximating Euclidean metrics by digital distances. Chapter 2 discusses digital distances, their classes, and hierarchies. In recent works, it has been shown that generalization of a family of distance functions is possible, and many of the results derived previously can be shown as special cases of the properties of the general class of distance functions. Chapter 3 presents analytical approaches for the analysis of errors of approximating Euclidean metrics by digital metrics. Chapter 4 considers the same with geometric approaches. In this regard, the properties of hyperspheres are also discussed. In Chap. 5 linear combinations of digital metrics for approximating Euclidean metrics are considered. Finally, in the concluding chapter, a few good digital distances in different dimensions of integral coordinate spaces are summarized. The chapter also concludes by highlighting a few open problems in this regard.

About four years back, I ventured into writing a monograph on this topic. But due to personal reasons, I could not proceed and a sort of inertia gripped me. I am thankful to the publisher, who showed interest in reviving the project and motivated me to complete it. I take this opportunity to express my gratitude to my supervisor Prof. B. N. Chatterji, who had provided immense support in my early research days, and encouraged me on exploring my independent research thoughts, even though they appeared simple and naive. My friend Prof. P. P. Das, who has contributed significantly to the development of the theory of digital distances, introduced me to this area of research. I always wondered how he could obtain all those amazing equations and expressions of digital metrics and their properties. It was always a pleasure to work with him and learn from him. My son Rudrabha is always curious about what his father is doing on the computer. He himself has become quite busy with his own research work. Even then, whenever needed, he provided all sorts of assistance in resolving technical glitches in composing this manuscript during this period. My wife Jhuma has to bear with my long hours of engagement before the computer terminals. As a doctor, she has all the worries and concerns about my health. I am fortunate to receive her love and care. I lost my mother five years back. I am quite unfortunate to miss her blessings on this occasion. From my early years, my parents instilled in me the dream of pursuing academic excellence. With deep gratitude, love, and respect, I dedicate this book to my father.

Kharagpur, India   Jayanta Mukhopadhyay
June 2020

Contents

About the Author

Dr. Jayanta Mukhopadhyay (Mukherjee) received his B.Tech., M.Tech., and Ph.D. degrees in Electronics and Electrical Communication Engineering from the Indian Institute of Technology (IIT), Kharagpur, in 1985, 1987, and 1990, respectively. He joined the faculty of the Department of Electronics and Electrical Communication Engineering at IIT, Kharagpur, in 1990, and later moved to the Department of Computer Science and Engineering where he is presently a Professor. He served as the Head of the Computer and Informatics Center at IIT, Kharagpur, from September, 2004 to July, 2007. He also served as the Head of the Department of Computer Science and Engineering and the School of Information and Technology from April, 2010 to March, 2013.

He was a Humboldt Research Fellow at the Technical University of Munich, in Germany, for one year in 2002. He has also held short-term visiting positions at the University of California, Santa Barbara, University of Southern California, and the National University of Singapore. His research interests are in image processing, pattern recognition, computer graphics, multimedia systems, and medical informatics. He has supervised 25 doctoral students and published more than 300 research papers in journals and conference proceedings in these areas. He has authored a book on "Image and Video Processing in the compressed domain", and co-authored a book on "Digital Geometry in Image Processing".

Dr. Mukhopadhyay is a senior member of the IEEE. He also holds life membership of various professional societies in his areas of expertise such Indian Association of Medical Informatics (IAMI), Telemedicine Society of India (TSI), Indian Unit of Pattern Recognition and Artificial Intelligence (IUPRAI, India). He has served as a member of technical program committees of several national and international conferences, and served as Program Co-Chairs of Indian Conference on Computer Vision, Graphics and Image Processing (ICVGIP) in 2000 and 2008. He also served as Program Chairs of the International Workshop on Recent Advances in Medical Informatics in 2013 and 2014. He is serving as a member of the editorial boards of Journal of Visual Communication and Image Representation, and Pattern Recognition Letters published by Elsevier.

He received the Young Scientist Award from the Indian National Science Academy in 1992, and is a Fellow of the Indian National Academy of Engineering (INAE).

Acronyms

ANAE	Average normalized absolute error
ARE	Average relative error
CMID	Chamfer mask induced distance
CWD	Chamfering weighted distance
EARE	Empirical average relative error
EARES	Empirical average relative error on sampling
ECM	Equivalent chamfer mask
EDT	Euclidean distance transform
EMNAE	Empirical maximum normalized absolute error
EMRE	Empirical maximum relative error
EMRES	Empirical maximum relative error on sampling
ENAE	Empirical normalized average error
ERB	Equivalent rational ball
ERM	Equivalent rational mask
HOD	Hyperoctagonal distance
LSE	Least squares estimation
LWD	Linear combination form of weighted distance
MAE	Maximum absolute error
MAT	Medial axis transform
MNAE	Maximum normalized absolute error
MRE	Maximum relative error
MSE	Mean squared error
NAE	Normalized absolute error
OEN	Overestimated norm
RMSE	Root mean squared error
UEN	Underestimated norm
UMSE	Unbiased mean squared error

WGNS   Weighted generalized neighborhood sequence
WtCWD   Linear combination of weighted t-cost and chamfering weighted
          distance
WtD   Weighted t-cost distance

Symbols and Notations

$\mathcal{R}$	Set of real numbers.
$\mathcal{P}$	Set of nonnegative real numbers.
$\mathcal{R}^+$	Set of positive real numbers.
$\mathcal{Z}$	Set of integers.
$\mathcal{N}$	Set of non-negative integers.
$\mathcal{Q}$	Set of rational numbers.
$\mathcal{Z}_M$	A set of integers: $\{0, 1, 2, \ldots, M\}$
$|S|$	The cardinality of set S.
$\mathcal{R}^n$	n-D real space.
$\mathcal{Z}^n$	n-D integral space.
$\mathcal{E}_S(x)$	Expectation of a random variable x over a set S.
$\bar{u}$	$\bar{u}$ is a point in $\mathcal{R}^n$ (or $\mathcal{Z}^n$), and also denoted as $\bar{u} = (u_1, u_2, \ldots, u_n)$, $\forall i$, $u_i \in \mathcal{R}$ (or $u_i \in \mathcal{Z}$).
$\bar{0}$	$(0, 0, \ldots, 0)$.
$\|\bar{u}\|$	Magnitude of $\bar{u}$, .i.e. $\sqrt{u_1^2 + u_2^2 + \cdots + u_n^2}$.
$n!$	Factorial of a nonnegative integer n.
$|x|$	Absolute value or magnitude of x, where x is a number.
$\lceil x \rceil$	Ceiling of x, i.e. the smallest integer greater than or equal to $x \in \mathcal{R}$.
$\lfloor x \rfloor$	Floor of x i.e. the greatest integer smaller than or equal to $x \in \mathcal{R}$.
$u_{(j)}$	jth maximum magnitude of components of $\bar{u}$.
$d(\bar{x}, \bar{y})$, $d(\bar{u})$	If $d(\bar{u})$ denotes a norm in a space, its induced distance function is denoted by $d(\bar{x}, \bar{y})$.
$DT(p)$	Distance transform at a point p.
$FT(p)$	Feature transform at a point p.
$MRE(\rho)$	Maximum relative error of a distance function ρ, with respect to the Euclidean metric in the same dimensional space.

$\Delta_{max}^{(n)}(\rho_1, \rho_2)$	Maximum absolute deviation between two norms ρ_1 and ρ_2 in a bounded region in n-D.
$E_{v\pi}^{(n)}(\rho)$	Volume-π error of a norm ρ in n-D.
$E_{s\pi}^{(n)}(\rho)$	Surface-π error of a norm ρ in n-D.
$E_{\psi\pi}^{(n)}(\rho)$	Shape-π error of a norm ρ in n-D.
κ_{opt}	Optimum scale for a distance function providing minimum MRE.
$E^{(n)}(\bar{u})$	Euclidean norm at $\bar{u} \in \mathscr{R}^n$.
$L_p(\bar{u})$	L_p norm at $\bar{u} \in \mathscr{R}^n$, i.e. $(\sum_{i=1}^{n} |u_i|^p)^{\frac{1}{p}}$.
type-m or $O(m)$-neighbor	$\bar{v}$ is a *type-m* or $O(m)$ neighbor if the coordinate values of at most m number of components of $\bar{v}$ differ by at most 1 from those of $\bar{u}$. This implies $\sum_{i=1}^{n} |u_i - v_i| \leq m$ and $\forall i, |u_i - v_i| \leq 1$. When equality condition holds the $\bar{v}$ is a *strict type-m* or a *strict $O(m)$* neighbor of $\bar{u}$.
B	B is a cyclic or periodic neighborhood sequence in n-D such that $B = \{b(1), b(2), \ldots, b(p)\}$, where $b(i) \in \{1, 2, \ldots, n\}$ is a neighborhood type. B is also used to denote the hyperoctagonal distance defined by it.
$CPERM(B)$	The set of cyclic permutations of the periodic neighborhood sequence B.
$[\omega_1, \omega_2, \ldots, \omega_n]$	A sorted neighborhood sequence B, with a fixed vector representation such that *type-i* neighborhood occurs ω_i times in the sequence.
$\mathbb{N}$	A *generalized neighborhood* of a point so that $\bar{v} \in \mathscr{L}^n$ is a neighbor of $\bar{u} \in \mathscr{L}^n$, if $\exists \bar{p} \in \mathbb{N}$, such that $\bar{v} = \bar{u} + \bar{p}$. $\mathbb{N}$ maintains 2^n-Symmetry.
$\mathbb{P}$	A *weighted generalized neighborhood* of a point given by $(\mathbb{N}, W(\cdot))$, where $W : \mathbb{N} \to \mathscr{R}^+$ to $\mathbb{N}$, such that W is centrally symmetric.
$\mathbb{B}$	A *weighted generalized neighborhood sequence* (WGNS) $\mathbb{B} = \{\mathbb{P}_1, \mathbb{P}_2, \ldots, \mathbb{P}_p\}$ with the length of period p, where $\mathbb{P}_i$ defines a weighted generalized neighborhood.
$d_4(\bar{u})$	City block or 4-neighbor norm at $\bar{u} \in \mathscr{L}^2$.
$d_8(\bar{u})$	Chess board or 8-neighbor norm at $\bar{u} \in \mathscr{L}^2$.
$d_{oct}(\bar{u})$	Octagonal norm at $\bar{u} \in \mathscr{L}^2$.
$d_6(\bar{u})$	6-neighbor norm at $\bar{u} \in \mathscr{L}^3$.
$d_{18}(\bar{u})$	18-neighbor norm at $\bar{u} \in \mathscr{L}^3$.
$d_{26}(\bar{u})$	26-neighbor norm at $\bar{u} \in \mathscr{L}^3$.

$d_m^{(n)}$	m-neighbor distance in $\mathscr{Z}^n$, $1 \leq m \leq n$.
$\delta_m^{(n)}$	Real m-neighbor distance in $\mathscr{R}^n$, $m \in \mathscr{R}$.
$d_B^{(n)}$	Hyperoctagonal distance defined by the neighborhood sequence B.
$d_B^{(2)}(\cdot\|q,p)$	A simple octagonal distance in $\mathscr{Z}^2$ such that $m = \frac{q}{p} \in [1, 2]$.
$D_t^{(n)}$	A t-cost norm in n-D, $1 \leq t \leq n$.
$WtD^{(n)}(\bar{u}; W)$	Weighted t-cost norm [42] in n-D with the ordered set of weights W, where the distance value is in $\mathscr{R}$.
$WtD_{sub}^{(n)}$	Simple upper bound optimized weighted t-cost distance in n-D.
$WtD_{isr}^{(n)}$	Inverse square root weighted t-cost distance in n-D.
$LWD^{(n)}(\bar{u}; \Gamma)$	Linear combination form of weighted distance (LWD) norm in n-D with the ordered set of weights Γ, where the distance value is in $\mathscr{R}$.
$CWD^{(n)}(\bar{u}; \Delta)$	Chamfering weighted distance (CWD) norm in n-D with the ordered set of weights Δ, where the distance value is in $\mathscr{R}$.
$CWD_{eu}^{(n)}$	The CWD in n-D with the set of weights $\{\sqrt{i}\|1 \leq i \leq n\}$.
$CWD_{euopt}^{(n)}$	The $CWD_{eu}^{(n)}$ scaled by an optimum scale κ_{opt} providing minimum MRE.
$<\delta_1, \delta_2, \ldots, \delta_n>$	The CWD with the ordered set of weights as $\{\delta_1, \delta_2, \ldots, \delta_n\}$.
$CWD_{3\_4\_gen}^{(n)}$	The CWD with the set of weights as $<3, 4, \ldots, n+2>$.
$CWD_{eu\_int}^{(n)}$	The CWD with a set of integral weights approximating $CWD_{eu}^{(n)}$ (Eq. (6.1)).
$WtCWD^{(n)}(\bar{u}; W, \Delta, a, b)$	A linear combination of $WtD^{(n)}(\bar{u}; W)$ and $CWD^{(n)}(\bar{u}; \Delta)$, such that $WtCWD^{(n)}(\bar{u}; W, \Delta, a, b) = a \cdot WtD^{(n)}(\bar{u}; W) + b \cdot CWD^{(n)}(\bar{u}, \Delta)$.
$WtCWD_{isr\_eu}^{(n)}$	A linear combination of $WtD_{isr}^{(n)}$ and $CWD_{eu}^{(n)}$.
$WtCWD_{sub\_rec}^{(n)}$	A linear combination of $WtD_{sub}^{(n)}$ and the CWD with reciprocal sets of weights.
$\mathscr{M}$	A chamfer mask.
$\mathscr{M}_g$	The generator of a chamfer mask.
$\mathscr{M}_{55}(a, b, c)$	A chamfer mask of size 5×5 in 2D, with parameters a, b and c.
$\mathscr{M}_{555}(a, b, c, d, e, f)$	A chamfer mask of size $5 \times 5 \times 5$ in 3D, with parameters a, b, c, d, e, and f.

$\mathcal{M}_\mathbb{B}$ The equivalent chamfer mask of a WGNS $\mathbb{B}$.

$\phi(\bar{u})$ The set of points in n-D formed by all possible signed permutation of coordinate values of $\bar{u}$, i.e. $\phi(\bar{u}) = \{\bar{v}|\bar{v}$ is formed by permuting $s_i u_i$s where s_i is either $+1$ or $-1\}$.

Chapter 1
Geometry, Space, and Metrics

More than two thousand years ago, Euclid outlined the analytical models of geometry of 2-D plane on the basis of his definitions of primitive geometrical entities such as points, straight lines, circles, angles, right-angles, etc., and five fundamental postulates on their existence and characterization, such as existence and uniqueness of a straight segment among two points, on straightness of segments between two points, formation of a right angle, existence of a circle of any arbitrary radius centering a point, and given a straight line existence of its parallel line through a non-collinear point. Euclid's postulates and their subsequent application to analyze the geometry of our space are so natural, that for more than 2000 years, mathematicians could not think about the existence of any other form of geometry. That is how from our early lessons of mathematics and geometry, we are more used to consider that the conventional geometry of the space around us is Euclidean. Strictly speaking, it is three dimensional Euclidean space. If we restrict the geometry in a plane (e.g., the floor of a building), it turns out to be two-dimensional Euclidean space. Again if a person is conservative enough to take into account the curvature of earth, the 2-D planar floor is not strictly Euclidean. It may be approximated to a greater accuracy using the concept of the Riemannian space. The latter space consists of the points lying on a spherical surface and the distance between any two points is computed by the length of the arc defined by the circle with the center and the radius as the same as those of the sphere.

A digital image, captured by an ordinary camera, is usually two-dimensional. Each element or pixel of the image has integral coordinate positions, contrary to the respective 2-D Euclidean space which is continuous and having infinite points in the neighborhood of any point in the space. This indicates the fact that the geometry in the image space is a non-Euclidean. We need to study this geometry and its approximation to the Euclidean world. In general, the digital image space is referred to as the digital space. Its dimension could be more than two, for example, digital images obtained from different medical imaging technologies such as CT-SCAN, MRI, PET, etc. In a digital space the points are represented by integral coordinate positions, for example, by row and column numbers of a 2-D image matrix. Hence,

J. Mukhopadhyay, *Approximation of Euclidean Metric by Digital Distances*,
https://doi.org/10.1007/978-981-15-9901-9_1

the geometry in this space is called *digital geometry*. But, this geometry is not unique. In different ways, digital geometry may be defined depending upon the neighborhood definition of a point in the space. It is necessary to understand the geometry in such a space in relation to our conventional understanding of Euclidean space. The present book is an endeavor to improve our understanding on this aspect, in particular, how digital distances and the geometry defined by them deviate from the properties of their Euclidean counterparts.

1.1 Metrics and Neighbors

The geometry in a space X is governed by the definition of the distance function between two points x and y. However, for a compact neighborhood definition, this distance function should satisfy the metricity property as defined below.

Definition 1.1 A *metric* on a non-empty set, X, is a function, $d : X \times X \to \mathscr{R}$, such that for all x, y, z in X, it satisfies the following conditions:

1. *Positive definiteness*: $d(x, y) \geq 0$, and $d(x, y) = 0$, if and only if, $x = y$,
2. *Symmetry*: $d(x, y) = d(y, x)$, and
3. *Triangular inequality*: $d(x, z) \leq d(x, y) + d(y, z)$.

Next, we define a norm in relation to a metric.

Definition 1.2 Let X be a vector space over $\mathscr{R}$. A function $g : X \to \mathscr{R}$ is called a *norm* on X, if for all $\bar{x}$, $\bar{y}$, $\bar{z}$ in X, it satisfies

1. *Positive definiteness*: $g(\bar{x}) > 0$, if $\bar{x} \neq \bar{0}$, and $g(\bar{0}) = 0$,
2. *Positive homogeneity*: $g(\lambda \bar{x}) = |\lambda| g(\bar{x})$, and
3. *Triangular inequality*: $g(\bar{x} + \bar{y}) \leq g(\bar{x}) + g(\bar{y})$.

For some metrics in a space, even though they are not strictly a norm, they asymptotically behave like a norm as the distance values become larger. We call them *asymptotic norms*.

Definition 1.3 A function $g : X \to \mathscr{R}$ in a vector space X is an *asymptotic norm*, if it satisfies the properties of positive definiteness, triangular inequality and the positive homogeneity asymptotically implying $\lim_{\lambda \to \infty} \frac{g(\lambda \bar{x})}{|\lambda| . g(\bar{x})} = 1$.

It may be noted that a norm function $g(.)$ also defines a metric $d(.)$ in $\mathscr{R}^n$, such that $d(\bar{x}, \bar{y}) = g(\bar{x} - \bar{y})$. In our discussion throughout this book, we represent the norm induced by a distance function $f(\bar{u}, \bar{v})$ by the same functional name $f(.)$, i.e., $f(\bar{w})$, where $\bar{w} = \bar{u} - \bar{v}$.

A metric is called *translation invariant*, if $d(\bar{x} + \bar{z}, \bar{y} + \bar{z}) = d(\bar{x}, \bar{y})$, $\forall \bar{x}, \bar{y}, \bar{z} \in X$. If a metric is translation invariant and also satisfies the property of positive homogeneity, it induces a norm.

The space X referred to here could be a 2-D or a 3-D integral coordinate space for digital images. We denote them, respectively, as $\mathscr{Z}^2$ and $\mathscr{Z}^3$, where $\mathscr{Z}$ is the set of integers. Similarly, for conventional Euclidean space, we denote the 2-D and 3-D coordinate spaces as $\mathscr{R}^2$ and $\mathscr{R}^3$, where, $\mathscr{R}$ denotes the set of real numbers.

One may easily check that the Euclidean distance functions in both 2-D ($\mathscr{R}^2$) and 3-D ($\mathscr{R}^3$) are metrics. In fact, the generalized Euclidean distance function in n-D ($\mathscr{R}^n$) is a metric. For the sake of convenience, we denote a point p in an n-D Cartesian coordinate space ($\mathscr{R}^n$ or $\mathscr{Z}^n$) by the notation $\bar{p} = (p_1, p_2, \ldots p_n)$. In that case, the Euclidean distance function $E^{(n)}(\bar{p}, \bar{q})$ is defined as follows:

$$E^{(n)}(\bar{p}, \bar{q}) = \sqrt{\sum_{i=1}^{n}(p_i - q_i)^2} \qquad (1.1)$$

Typically in 2-D ($\mathscr{R}^2$) and 3-D ($\mathscr{R}^3$), these functions are expressed as

$$E^{(2)}(\bar{p}, \bar{q}) = \sqrt{(p_1 - q_1)^2 + (p_2 - q_2)^2} \qquad (1.2)$$

$$E^{(3)}(\bar{p}, \bar{q}) = \sqrt{(p_1 - q_1)^2 + (p_2 - q_2)^2 + (p_3 - q_3)^2} \qquad (1.3)$$

In the Euclidean space, there are infinite number of points in the neighborhood of any point. The distance between two neighbors also should be infinitesimally small and in the limiting case it approaches to zero. In that case, we define the *ε-neighborhood of a point* (refer to Fig. 1.1) as follows:

Definition 1.4 Two points $\bar{p}$ and $\bar{q}$ in $\mathscr{R}^n$ are said to be *ε-neighbor*, if $E^{(n)}(\bar{p}, \bar{q}) \leq \varepsilon$.

Interestingly, the Euclidean distance function is also a metric in the digital space such as in $\mathscr{Z}^2$ (for 2-D) and $\mathscr{Z}^3$ (for 3-D). However, the neighborhood of a point in that case is not infinite. Another problem of using Euclidean distance function is

Fig. 1.1 ε-neighborhood of a point

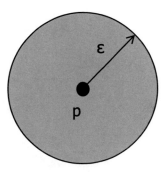

that, from any point, at any distance, there exists a point in the space which is not true for the digital space. In this case, as the coordinate space is integral, only for discrete distance values, this is true. It does not mean that for any non-integral value, there does not exist any point in the digital space. It may or may not. For example, from the point $(0, 0)$ (in Z^2) there exists a point at an Euclidean distance of $\sqrt{5}$ (e.g., $(1, 2)$). That is why digital distances are preferred to it in the digital space. In *true digital distances*, distances between integral coordinate points are also integers. Additionally, if these functions are metrics, the neighborhood of any point remains bounded, and the number of points in the neighborhood is also finite.

Definition 1.5 A *true digital distance function* is a metric in $\mathscr{Z}^n$ and it is of the form $\mathscr{Z}^n \times \mathscr{Z}^n \to \mathscr{Z}$.

Some of the popular distance functions in 2-D and 3-D are given below [65].
In 2-D:

1. City block (or 4-neighbor) distance: $d_4(\bar{i}, \bar{j}) = \mid i_1 - j_1 \mid + \mid i_2 - j_2 \mid$
2. Chess board (or 8-neighbor) distance: $d_8(\bar{i}, \bar{j}) = \max(\mid i_1 - j_1 \mid, \mid i_2 - j_2 \mid)$.

In 3-D

1. 6-neighbor distance: $d_6(\bar{i}, \bar{j}) = \mid i_1 - j_1 \mid + \mid i_2 - j_2 \mid + \mid i_3 - j_3 \mid$
2. 18-neighbor distance: $d_{18}(\bar{i}, \bar{j}) = \max(\lceil \frac{d_6(\bar{i}, \bar{j})}{2} \rceil, d_{26}(\bar{i}, \bar{j}))$
3. 26-neighbor distance: $d_{26}(\bar{i}, \bar{j}) = \max(\mid i_1 - j_1 \mid, \mid i_2 - j_2 \mid, \mid i_3 - j_3 \mid)$.

It may be checked, that all the above distance functions are metrics and true digital functions. The names of these distance functions are also tagged with the number of neighbors of a point in the digital space [63]. In Fig. 1.2, the neighboring points (shown in gray color) of a point (in black color) in a 2-D discrete grid are shown.

As the distance values are positive integers (including zero), the neighbors of a point are defined as follows:

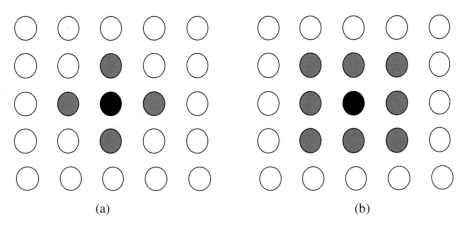

(a)   (b)

Fig. 1.2 **a** 4-neighbors of a point and **b** 8-neighbors of a point in $\mathscr{Z}^2$

Definition 1.6 For true digital distances, two points are neighbors to each other if the distance between them is exactly 1, which is the minimum distance between two distinct points in the space.

There is also another type of distance function, which is defined in the integral coordinate space, but whose values may not be always integers. They could be real numbers. However, their neighbors are placed at discrete distance values only. This type of distance function, we call *semi-digital distance function*.

Definition 1.7 A *semi-digital distance function* is a metric in $\mathscr{Z}^n$ and it is of the form $\mathscr{Z}^n \times \mathscr{Z}^n \to \mathscr{R}$.

An example of a semi-digital distance function is a *weighted distance function* [6], where distances of a set of neighboring points are given by a set of weights (see Sect. 2.3 for full definition and discussion). If the weights are suitably chosen, the function may satisfy the properties for becoming a metric. The definition of neighbors for a semi-digital distance function is not as simple as that of a true digital distance function. In this case, a neighbor could be at a varying distance, one of a set of discrete distance values. We note that all the digital distances mentioned above are also metric in the real spaces ($\mathscr{R}^2$ or $\mathscr{R}^3$). However, the distance values are real numbers in these cases. For these metrics also, we may define the ε-neighborhood as defined in Definition 1.4. The structures of these ε-neighborhoods in 2-D are shown in Fig. 1.3.

We could easily observe that for digital distances, these structures are exactly the convex hull formed by the digital neighbors. For Euclidean metric, however, this is not true. In this case, the neighborhood structure in $\mathscr{R}^2$ is circular, while the convex hull formed by the neighbors in $\mathscr{Z}^2$ (with Euclidean distance function) is of the shape of a diamond (as same as with the city block (or 4-neighbor) distance function).

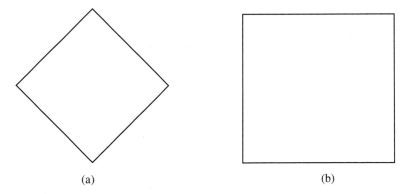

(a)　　　　　　　　　　　　　　　(b)

Fig. 1.3 ε-neighborhood of digital distances when extended to continuum in real space ($\mathscr{R}^2$): **a** 4-neighbor distance, and **b** 8-neighbor distance

1.1.1 Geometry of Disks

The notion of neighborhood is extended to define disks of a distance function.

Definition 1.8 A *disk* $C(q, b)$ in a space X and the distance function d is defined as the set of points such that their distances from $q \in X$ are less than or equal to b (which is a scalar value, a nonnegative real number or integer). q is the center of the disk and b is its radius. Formally, the disk is defined as follows:

$$C(q, b) = \{p \mid (p \in X) \text{ and } (d(p, q) \leq b)\}$$

In 2-D Euclidean space, the shape of a disk is circular, whereas in 3-D its shape is spherical (Fig. 1.4). However, it is interesting to observe the shapes of the disks in the digital space with digital distance functions. In 2-D, the shape is a convex polygon and in 3-D, it is a convex polyhedron. Some of the examples are shown in Fig. 1.5. In n-D, a disk is also called a *hypersphere*.

There is a very simple but powerful characterization of disks by which we can establish whether a distance function is a norm [4].

Theorem 1.1 *A distance function is a norm if and only if its disk (or hypersphere) is convex, symmetric and homogeneous.*

1.1.2 Paths and Distances

Definition 1.9 A *path* between two points p and q is defined as the sequence of points $x_0(= p), x_1, x_2, x_3, \ldots, x_n(= q)$ such that the consecutive pair of points x_i and x_{i+1} are neighbors.

In Fig. 1.6, an example of a path formed from the definition of ε-neighborhood in 2-D Euclidean space is illustrated. The curvilinear path is formed as $\varepsilon \to 0$. In discrete space $\mathscr{Z}^n$, the path is formed by consecutive neighbors and the number of points between two end points are countable. If the path in the integral space is

Fig. 1.4 Disks of Euclidean metrics: **a** 2-D: Circle and **b** 3-D: Sphere

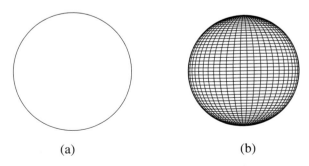

(a)   (b)

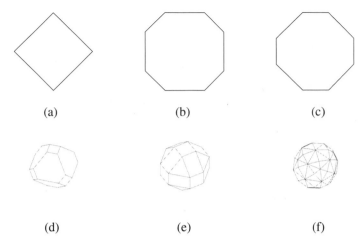

(a)   (b)   (c)

(d)   (e)   (f)

Fig. 1.5 Examples of disks of digital distances: **a–c** in 2-D and **d–f** in 3-D

Fig. 1.6 An illustration of a path formed by ε-neighborhood of points

formed by n points, the length of this path is the sum of distances between consecutive neighbors. For example, for the four-neighbor distance in this case the length would be $n - 1$. From the length of the path also, the distance between two points can be defined.

Definition 1.10 The distance between two points is defined as the *length of the shortest path between them.*

In Euclidean space this path is the straight line between two points. However, for digital distances, there could be more than one minimal path. An example is shown in Fig. 1.7.

1.1.3 Distance Transform

Distance transform was introduced by Rosenfeld and Pfaltz in the sixties of the last century [64]. It finds applications in shape analysis and many other image processing operations, such as, skeletonization [64, 84], morphological operations, medial axis transform [5], decomposition of 3-D objects [78], object representation [68], computing shape descriptors [66], motion planning, object tracking [33, 37], etc. Distance transform of a set of points X (denoted as foreground) with reference to another set of points B (denoted as background) is defined as follows:

Fig. 1.7 Examples of the shortest paths between the two points in 2-D digital space with city block (or four-neighbor) distance function. Both the paths shown by solid and dashed lines are of the same length, which are of minimal lengths of all possible paths between these two points

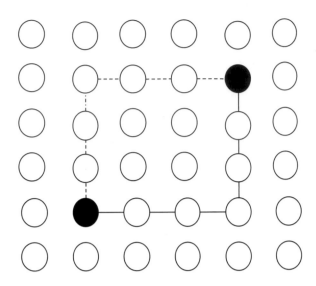

Definition 1.11 At every point $p \in X$, distance transform $DT(p)$ is its minimum distance from B, i.e.

$$DT(p) = \min_{q \in B}\{d(p,q)\}$$

On the other hand, the point q of B, which is at the minimum distance from p of X, is called its feature transform, denoted by $FT(p)$.

Definition 1.12 At every point $p \in X$, feature transform $FT(p)$ is the point in B which is at the least distance from p, i.e.

$$FT(p) = \arg\min_{q \in B}\{d(p,q)\}$$

$FT(p)$ is not unique at a given point p.

1.1.3.1 Medial Axis Transform (MAT)

One of the applications of distance transform is to compute medial axis transform (MAT) [5], of an object in an image. In this transform centers and radii of maximal disks contained within the object are computed. A maximal disk is defined as follows:

Definition 1.13 A disk is *maximal* in an object if it is not covered by any other disk, contained within the object.

Definition 1.14 *Medial axis transform* (MAT) of an object is the set of maximal disks contained in it.

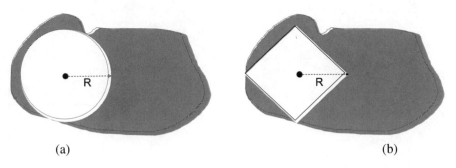

Fig. 1.8 A maximal disk in 2-D within the object **a** (Circle) of Euclidean metric and **b** (Diamond) of City block distance function

Given a disk with its center and radius, it is trivial to compute the set of object points covered by it. The union of all such sets formed from the disks of a MAT fully reconstructs the set of object points. Different applications of MAT in computing transformation of an object [39], normals at boundaries of 2-D objects in $\mathscr{Z}^2$ [49], cross-sections of 3-D objects in $\mathscr{Z}^2$ [48], etc., have been reported in the literature. It is also used for skeletonization of objects [40]. All these computations are sensitive to the shape of a maximal disk, which otherwise depends upon the chosen metric. In Fig. 1.8, examples of disks of Euclidean and city block distances are given. A maximal disk can be computed from the distance transform of an object. It is centered at a local maximum of the transform and the distance value provides the radius of the disk. As there are various advantages to deal with the perfect circular shape of Euclidean metric, the distance transform of Euclidean metric is preferred in shape analysis. However computation of exact Euclidean distance transform (EDT) is non-trivial, though there exists a linear time algorithms ($O(n^d)$) for a d-dimensional binary image with n^d grid points [16]. In another algorithm [81], the computation is performed in $O(dn^d)$. In [60], a four pass scanning algorithm is presented, where scanning could be made in any order to compute the EDT. There were also attempts to develop efficient algorithms for obtaining approximate EDT [58].

1.1.3.2 Distance Transform Through Chamfering

In digital space, digital distances are quite naturally defined and computation of distance transforms using those distances is relatively easy as they define neighborhoods of a finite sets of points. There exist various efficient algorithms for many of these distance transforms. A very popular algorithm is known as *chamfering* [6], which computes distances at every point by forward and backward scanning of the lattice of points from one end to the other end using a neighborhood mask, called *chamfer mask*. A chamfer mask defines the distances of neighbors of a point in a neighborhood grid. For example, in 2-D, with 4-neighbor definitions, the chamfer mask is defined as given in Fig. 1.9a. In the forward scan, the center of the mask is placed

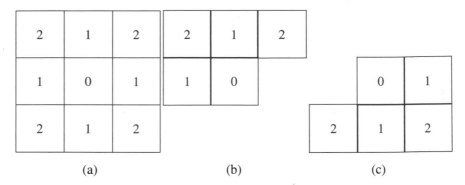

Fig. 1.9 Chamfer masks of: **a** 4-neighbor distance, **b** Mask for forward scan, and **c** Mask for backward scan

at every pixel and the distance value of the pixel (from the background) is updated by observing distances of its neighboring pixels, visited before. The order of visit of pixels follow a scanning from left to right and top to bottom of the image, and it uses upper triangular half of the mask as shown in Fig. 1.9b. In the backward scan, the order of visit takes place from right to left and bottom to top using the lower triangular half of the mask as shown in Fig. 1.9c. The algorithm runs in linear time. The same technique is extended to a higher dimension. For example, in 3-D, in each scan, the order of visit needs to accommodate the added dimension. So in this case, in forward scan, it involves an ordering from left to right, top to bottom, and front to back, and in the backward scan, the reverse order is followed. A brief summary of different approaches of chamfering is available in [17], where an efficient algorithm using separable distance transformation of path-based metrics in n-D has been discussed.

1.1.3.3 Computation of Local Maxima of the Distance Transform

The other aspect of computation of a maximal disk is to consider a suitable neighborhood around a point for finding a local maximum in the distance transform. Distance values of neighbors of a point are to be examined to check whether a local maximum occurs at that point. This is also another problem with the Euclidean metric. This checking is nontrivial and may require an explicit test of finding out whether a disk of a radius given by its distance value in the neighborhood covers the disk at that point p with a radius $DT(p)$ [56]. For many digital distances, it is as simple as finding local maxima around a finite neighborhood from the distribution of distance transform values in it [1]. All these make digital distances popular choices for computing distance transforms, which are used in shape analysis [66], and many other applications. However, if a distance function is close to the Euclidean metric, its properties may be approximated within a tolerable limit. This is one of the prime motivations for looking for a digital distance function close to the Euclidean metric.

1.2 Motivations for Finding Digital Distances Close to Euclidean Metrics

In the last section, we discuss why digital distances close to Euclidean metrics are preferred in applications involving distance transform and MAT. In this case, desirable properties of a disk of a MAT should follow those of an Euclidean disk. In this section, let us consider a few other contexts for searching digital distances to approximate Euclidean distance functions.

1.2.1 Reducing Computational Cost

In some applications, it is required to compute Euclidean metrics for a large number of cases. For example, to perform vector median filtering [2, 3], distances between a pair of vectors are computed at $O(N^2)$ for N number of vectors. As Euclidean metric requires square root operation, the cost of computation is higher compared to other distances, where only additions and comparisons are used, e.g., in computation of four-neighbor or eight-neighbor distances in 2-D integral coordinate space. However, at some points in the space, these two functions have high errors with respect to the Euclidean distance value. Under this scenario, a distance function having low error with Euclidean metric at every point in the space, yet computationally less costly, is preferred [67].

1.2.2 Discrete Nature of the Space

When we work within a discrete integral coordinate or digital space, it is often required to perform the Euclidean measurements, such as computing area, perimeter, volume, surface area, etc., in the digital metric space. The errors of approximation in digital space would be less if a suitable metric having low error with the corresponding Euclidean metric is used.

1.2.3 An Interesting Mathematical Problem

In mathematics, not every concept has immediate application, yet it is always interesting to look for new mathematical findings and solve challenging problems. From that aspect, finding metrics in discrete spaces, and relating geometry of those spaces with Euclidean space is of interest to many researchers. It is difficult to say whether those efforts are driven by the necessity of a useful application or just for fun to meet naive curiosity nourishing intellectual discourses!

1.3   Concluding Remarks

In this chapter, we review definitions of a distance function, a metric, a norm, and a hypersphere (disk) in an n-D space. We also discuss the relevance of studying distance functions, which closely approximate respective Euclidean metrics in their spaces. To compare proximity of different distance functions to Euclidean metrics, there exist several measures of errors in the literature. Broadly they could be categorized as *analytical measures* and *geometric measures*. In the analytical measures, the deviation of a distance value from the respective Euclidean distance value is considered, while in geometric measures, it is the deviation of geometry of a disk or a hypersphere of a distance function from that of the corresponding Euclidean metric taken into account. The details of different types of errors and their uses in studying Euclidean approximation of a distance function are discussed in subsequent chapters.

Chapter 2
Digital Distances: Classes and Hierarchies

In this chapter, we discuss different classes of digital distances. We consider distance functions in the generalized n-dimensional digital space ($\mathscr{Z}^n$), followed by their special cases in 2-D and 3-D spaces. Some of these classes could be related by a hierarchical relationship, where more general distance functions are placed at the top of hierarchies and their special cases are identified as sub-classes. Reviews of these distance functions are also available in [18, 50]. Though all the distance functions are defined in $\mathscr{Z}^n$, their definitions can trivially be extended in $\mathscr{R}^n$, such that if a function is a metric or a norm in $\mathscr{Z}^n$, the same also holds in $\mathscr{R}^n$. We begin with a class of distance functions named *m-neighbor distances*.

2.1 m-Neighbor Distances

m-neighbor distances are generalization of cityblock and chessboard distances in 2-D to any arbitrary n-dimension. Das et al. [24] proposed extension of these functions in any arbitrary dimension. In $\mathscr{Z}^n$, at a point $\bar{u}$, n different types of neighborhoods are defined. We refer to them as $O(m)$, $1 \leq m \leq n$ neighborhoods and they are defined as

Definition 2.1 A point $\bar{v} \in \mathscr{Z}^n$ is said to be $O(m)$ (or *type-m*) neighbor of a point $\bar{u} \in Z^n$, if it satisfies the following conditions:

1. $0 \leq | u_i - v_i | \leq 1$, $1 \leq i \leq n$
2. $\sum\limits_{i=1}^{n} | u_i - v_i | \leq m.$

If the equality holds in the second condition above, we call $\bar{v}$ *strict* $O(m)$ (or *strict type-m*) neighbor of $\bar{u}$.

© The Author(s), under exclusive license to Springer Nature Singapore Pte Ltd. 2020
J. Mukhopadhyay, *Approximation of Euclidean Metric by Digital Distances*,
https://doi.org/10.1007/978-981-15-9901-9_2

In 2-D, $O(1)$ and $O(2)$ neighbors are 4-neighbors with the city block distance function and 8-neighbors with the chessboard distance function, respectively. Given a point $\bar{u} \in \mathcal{Z}^2$, its *strict* $O(2)$ neighbors are $\{(u_1 \pm 1, u_2 \pm 1)\}$. We also refer to $O(1)$ and $O(2)$ neighbors as *type*-1 and *type*-2 neighbors.

Likewise in 3-D, there exist 3 types of neighbors, namely $O(1)$ (or *type*-1), $O(2)$ (or *type*-2), and $O(3)$ (or *type*-3), respectively. They are also called 6-neighbors, 18-neighbors and 26-neighbors as defined by their distance functions (refer to Sect. 1.1).

As a generalization in n-D, we define n different types of neighborhoods (Definition 2.1). For each *type*-m (or $O(m)$) neighborhood definition, there is a distance function associated with it, which we refer to as the *m-neighbor distance* function. Its closed form expression is given in the following theorem. The proof is available in [24].

Theorem 2.1 *The m-neighbor distance function between two points $\bar{u}$ and $\bar{v}$ in $\mathcal{Z}^n$ is given by*

$$d_m^{(n)}(\bar{u}, \bar{v}) = \max \left\{ \max_{1 \le i \le n} \{| u_i - v_i |\}, \left\lceil \frac{\sum_{i=1}^{n} | u_i - v_i |}{m} \right\rceil \right\} \tag{2.1}$$

For *type*-1 and *type*-n neighbors, the distance functions assume the following simple forms:

$$d_1^{(n)}(\bar{u}, \bar{v}) = \sum_{i=1}^{n} | u_i - v_i |$$

$$d_n^{(n)}(\bar{u}, \bar{v}) = \max_{1 \le i \le n} \{| u_i - v_i |\}$$

In 2-D, *type*-1 neighbors define the city block distance function, whereas *type*-2 neighbors correspond to the chess board distance function. Similarly, in 3-D *type*-1 and *type*-3, neighbors define, respectively, 6-neighbor and 26-neighbor distance functions. The 18-neighbor distance function as described in the previous chapter (Sect. 1.1) is given by the distance function corresponding to the *type*-2 neighborhood definition in 3-D. In Table 2.1, a list of these functions is provided.

2.2   t-Cost Distances

Another class of distance functions called the *t-cost distance functions* is introduced in [31]. In this class of distance functions, two points in $\mathcal{Z}^n$ are defined as neighbors, when their corresponding hypervoxels share a hyperplane of any dimension. However, the cost associated with them could be at most t, $1 \le t \le n$, such that if

Table 2.1 m-neighbor distances in 2-D and 3-D

m	n	$d_m^{(n)}(\cdot)$	Closed form expressions
1	2	$d_1^{(2)}(\bar{u},\bar{v})$	$\mid u_1 - v_1 \mid + \mid u_2 - v_2 \mid$
2	2	$d_2^{(2)}(\bar{u},\bar{v})$	$\max(\mid u_1 - v_1 \mid, \mid u_2 - v_2 \mid)$
1	3	$d_1^{(3)}(\bar{u},\bar{v})$	$\mid u_1 - v_1 \mid + \mid u_2 - v_2 \mid + \mid u_3 - v_3 \mid$
2	3	$d_2^{(3)}(\bar{u},\bar{v})$	$\max\{\max(\mid u_1 - v_1 \mid, \mid u_2 - v_2 \mid), \mid u_3 - v_3 \mid), \lceil \frac{\mid u_1 - v_1 \mid + \mid u_2 - v_2 \mid + \mid u_3 - v_3 \mid}{2} \rceil\}$
3	3	$d_3^{(3)}(\bar{u},\bar{v})$	$\max(\mid u_1 - v_1 \mid, \mid u_2 - v_2 \mid), \mid u_3 - v_3 \mid)$

Table 2.2 t-cost distance functions in 2-D and 3-D

n	t	$D_t^{(n)}(\bar{u})$
2	1	$\max(\mid u_1 \mid, \mid u_2 \mid)$
	2	$\mid u_1 \mid + \mid u_2 \mid$
3	1	$\max(\mid u_1 \mid, \mid u_2 \mid, \mid u_3 \mid)$
	2	$\max(\mid u_1 \mid + \mid u_2 \mid, \mid u_2 \mid + \mid u_3 \mid, \mid u_3 \mid + \mid u_1 \mid)$
	3	$\mid u_1 \mid + \mid u_2 \mid + \mid u_3 \mid$

two consecutive points on the shortest path share a hyperplane of dimension r, the distance between them is taken as $\min(t, n - r)$. There are n distinct t-cost norms. They are defined as follows:

Definition 2.2

$$D_t^{(n)}(\bar{u}) = \sum_{i=1}^{t} u_{(i)} \text{ for } 1 \le t \le n$$

where $u_{(i)}$ is the ith maximum magnitude of a coordinate of the point $\bar{u} \in \mathscr{L}^n$.

In [31], it is shown that these distance functions are norms in $\mathscr{L}^n$ (as well as in $\mathscr{R}^n$ when $\bar{u} \in \mathscr{R}^n$). In Table 2.2, t-cost norms in 2-D and 3-D are illustrated.

In 2-D these distances are no different than the m-neighbor distances. The $D_1^{(2)}$ is the same as $d_2^{(2)}$ (or d_8 in Sect. 1.1) and both the $D_2^{(2)}$ and $d_1^{(2)}$ (or d_4 in Sect. 1.1) are of the same closed form expressions.

In 3-D also $D_1^{(3)}$ and $D_3^{(3)}$ correspond to $d_3^{(3)}$ (or d_{26} in Sect. 1.1) and $d_1^{(3)}$ (or d_6 in Sect. 1.1), respectively. But, $D_2^{(3)}$ is not the same as $d_2^{(3)}$ (or d_{18} in Sect. 1.1).

2.3 Weighted Distances

Both m-neighbor and t-cost distances can be generalized by providing a flexible nonuniform cost assignment to the neighbors of a point in $\mathscr{L}^n$. We refer to these assigned costs as weights. We may assign varying weights for measuring the distance between a pair of *strict* $O(i)$ neighbors. Some of them also satisfy the property of a metric. This class of distance function is called *weighted distances*. In 1984, Borgefors [6], proposed these distances in 2-D considering the nonuniform cost for different types of motions (such as vertical/horizontal motions and diagonal motions in 2-D) by extending the concept of "chamfer-Euclidean" distance discussed in [41]. If the distance value is restricted to be an integer, these weights are integers. A typical functional form of a weighted norm with weights a and b for $O(1)$ and strict $O(2)$ neighbors in 2-D are given below.

In 2-D:

$$d^{(2)}_{<a,b>}(\bar{u}) = au_{(1)} + (b-a)u_{(2)}$$

where a and b are positive weights and $a \leq b \leq 2a$.

In 3-D:

$$d^{(3)}_{<a,b,c>}(\bar{u}) = au_{(1)} + (b-a)u_{(2)} + (c-b)u_{(3)}$$

where a, b and c are positive weights corresponding to $O(1)$, strict $O(2)$ and strict $O(3)$ neighbors, respectively, and $b \leq 2a$, $b \leq c$ and $a + c \leq 2b$. It should be noted that in the above $u_{(i)}$ denotes the i th maximum magnitude of a coordinate of the point $\bar{u}$.

In [43], two different analytical forms of weighted distances in n-D are discussed. Both representational forms are equivalent to each other.

Definition 2.3 The *linear combination form of weighted distance* (LWD) [2] is defined as

$$LWD^{(n)}(\bar{u}; \varGamma) = \sum_{i=1}^{n} \gamma_i u_{(i)} \qquad (2.2)$$

where $\varGamma = \{\gamma_i | \text{for some } k, \gamma_k > 0 \text{ and, } \gamma_i \geq 0, \ i = 1 \dots n, i \neq k\}$ and $u_{(i)}$ is the ith maximum magnitude of components of $\bar{u} \in \mathscr{L}^n$.

The following theorem states the necessary and sufficient conditions for an LWD becoming a norm. The proof is available in [2].

Theorem 2.2 $LWD^{(n)}(\bar{u}; \varGamma)$ *is a norm, if and only if*

$$\gamma_1 > 0, \ \text{ and, } \ \gamma_1 \geq \gamma_2 \geq \dots \geq \gamma_n \geq 0 \qquad (2.3)$$

The representation of a weighted distance as given in Definition 2.3, is also provided in an alternative form [43].

Definition 2.4 An equivalent representation of weighted distance as defined in Definition 2.3, is given by

$$CWD^{(n)}(\bar{u}; \varDelta) = \delta_1 u_{(1)} + \sum_{i=2}^{n}(\delta_i - \delta_{i-1})u_{(i)} \qquad (2.4)$$

where $\Delta = \{\delta_i | \text{for some } k, \delta_k > 0 \text{ and}, \delta_i \geq 0, i = 1 \ldots n, i \neq k\}$ and $u_{(i)}$ is the ith maximum magnitude of coordinates of $\bar{u} \in \mathscr{L}^n$.

The advantage of the above definition is that it provides the chamfering mask for computing a distance transform from its set of weights, if it is a norm.[1] In that case, δ_i is the distance of the strict ith m-neighbor [24] of a point for $i = 1, 2, \ldots, n$. In [7, 10, 11], the corresponding expressions of 2-D, 3-D, and 4-D, respectively, are provided. As Definition 2.4 is associated with a chamfering mask, the distance function in this form is called the *chamfering weighted distance* (CWD). Following conventional form of its representation in the literature, let us denote a CWD with an ordered set of $\Delta = \{\delta_1, \delta_2, \ldots, \delta_n\}$, as $< \delta_1, \delta_2, \ldots, \delta_n >$. It may be noted that the relationship of weights of a $CWD^{(n)}(\bar{u}; \Delta)$ and those of its corresponding equivalent $LWD^{(n)}(\bar{u}; \Gamma)$ is expressed in the following form:

$$\delta_i = \sum_{j=1}^{i} \gamma_j \text{ for } 1 \leq i \leq n \tag{2.5}$$

If the CWD is not a norm, it should not be associated with the corresponding chamfering mask constructed from its weights. Even if such a mask defines a metric in the space, the expression of the corresponding distance function would be different from a CWD (refer to Eq. (2.4)). There are examples [10] of such chamfer distances not in the form $CWD^{(n)}(\bar{u}; \Delta)$. In 3D when $\delta_1 + \delta_3 > 2\delta_2$, the functional form of the metric is different from the CWD, and it is not equivalent to the class of LWD. There are general forms of chamfer distances discussed in [35]. In this book, we restrict ourselves to their definitions in the forms of LWD (Definition 2.3) or CWD (Definition 2.4).

From Theorem 2.2, it implies that a $CWD^{(n)}(\bar{u}; \Delta)$ is a norm if it satisfies the following:

$$\delta_1 \geq (\delta_2 - \delta_1) \geq \cdots \geq (\delta_n - \delta_{n-1}) \geq 0 \tag{2.6}$$

The above equation (Eq. (2.6)), can be rewritten as the following:

$$\delta_{i-1} + \delta_{i+1} \leq 2\delta_i \tag{2.7}$$

and,

$$\delta_i \geq \delta_{i-1} \tag{2.8}$$

---

[1] We should also note that this definition is a simplification of the one given in [35] (refer to Theorem 2.3 of [35]), for chamfer masks made only of canonical basis vectors and can also be inferred from the latter.

Above two equations are used to state the following theorem:

Theorem 2.3 *If $CWD^{(n)}(\bar{u}; \Delta)$ is a norm, then*

$$
\begin{aligned}
\delta_1 &\leq \delta_2 < 2\delta_1, \\
\delta_2 &\leq \delta_3 < \tfrac{3}{2}\delta_2, \\
\delta_3 &\leq \delta_4 < \tfrac{4}{3}\delta_3, \\
&\;\;\vdots \\
\delta_{n-1} &\leq \delta_n < \tfrac{n}{n-1}\delta_{n-1}
\end{aligned}
\tag{2.9}
$$

Proof First we show that the conditions are satisfied for $n = 2$, and 3. Next, we use induction to prove the theorem.

For $n = 2$,

$$
\delta_1 \geq \delta_2 - \delta_1 \geq 0 \tag{2.10}
$$

As $\delta_2 - \delta_1 \geq 0$, $\delta_2 \geq \delta_1$.
Again

$$
\begin{aligned}
&\delta_1 \geq \delta_2 - \delta_1 \\
\Rightarrow\; &\delta_2 \leq 2\delta_1
\end{aligned}
\tag{2.11}
$$

Hence, it holds for $n = 2$.

For $n = 3$

$$
\delta_1 \geq \delta_2 - \delta_1 \geq \delta_3 - \delta_2 \geq 0 \tag{2.12}
$$

As $\delta_3 - \delta_2 \geq 0$ and $\delta_2 - \delta_1 \geq 0$, $\delta_3 \geq \delta_2 \geq \delta_1 \geq 0$.
Again

$$
\begin{aligned}
&\delta_2 - \delta_1 \geq \delta_3 - \delta_2 \\
\Rightarrow\; &\delta_3 \quad \leq 2\delta_2 - \delta_1 \\
\Rightarrow\; &\delta_3 \quad \leq 2\delta_2 - \tfrac{\delta_2}{2} \quad \text{from Eq. (2.11)} \\
\Rightarrow\; &\delta_3 \quad \leq \tfrac{3}{2}\delta_2
\end{aligned}
\tag{2.13}
$$

Hence, it holds for $n = 3$.

Suppose the condition is true for $n = m - 1$. We need to show that it also holds for $n = m$. This is shown below.

As $\delta_1 \leq \delta_2 \leq \ldots \delta_{m-1}$, and $(\delta_m - \delta_{m-1}) \geq 0$, $\delta_1 \leq \delta_2 \leq \ldots \delta_{m-1} \leq \delta_m$.
Again

$$
\begin{aligned}
&\delta_{m-1} - \delta_{m-2} \geq \delta_m - \delta_{m-1} \\
\Rightarrow\; &\delta_m \quad \leq 2\delta_{m-1} - \delta_{m-2} \\
\Rightarrow\; &\delta_m \quad \leq 2\delta_{m-1} - \tfrac{m-2}{m-1}\delta_{m-1} \quad \text{as } \delta_{m-1} \leq \tfrac{m-1}{m-2}\delta_{m-2} \text{ from induction.} \\
\Rightarrow\; &\delta_m \quad \leq \left(2 - \tfrac{m-2}{m-1}\right)\delta_{m-1} \quad = \tfrac{m}{m-1}\delta_{m-1}
\end{aligned}
\tag{2.14}
$$

Hence the theorem.   □

Table 2.3 Typical examples of weighted distances

Dimension	Δ (CWD)	Γ (LWD)	Notation/Name	Remarks and scaling to be used for approximating to Euclidean norm
2	$\{3, 4\}$	$\{3, 1\}$	$< 3, 4 >$	$\frac{1}{3}$
2	$\{0.95509, 1.36930\}$	$\{0.95509, 0.4142\}$	$CWD_{opt}^{(2)}$	Reported to be optimum for a performance measure [7]
2	$\{1, 1.35070\}$	$\{1, 0.35070\}$	$CWD_{opt*}^{(2)}$	Reported to be optimum with $\delta_1 = 1$ for a performance measure [7]
2	$\{0.9481, 1.3408\}$	$\{0.9481, 0.3927\}$	$CWD_{umse}^{(2)}$	Reported to have the minimum unbiased mean squared error (UMSE) in 2D [80]
3	$\{3, 4, 5\}$	$\{3, 1, 1\}$	$< 3, 4, 5 >$	$\frac{1}{3}$
3	$\{0.92644, 1.34065, 1.65849\}$	$\{0.92644, 0, 4142, 0.3178\}$	$CWD_{opt}^{(3)}$	Reported to be optimum for a performance measure [10]
3	$\{1, 1.31402, 1.62803\}$	$\{1, 0.31402, 0.3140\}$	$CWD_{opt*}^{(3)}$	Reported to be optimum with $\delta_1 = 1$ for a performance measure [10]
3	$\{0.894, 1.3409, 1.5879\}$	$\{0.894, 0.4469, 0.2470\}$	$CWD_{umse}^{(3)}$	Reported to have the minimum unbiased mean squared error (UMSE) in 3D [80]
4	$\{3, 4, 5, 6\}$	$\{3, 1, 1, 1\}$	$< 3, 4, 5, 6 >$	$\frac{1}{3}$
4	$\{0.9048, 1.3191, 1.6369, 1.9048\}$	$\{0.9048, 0.4143, 0.3178, 0.2679\}$	$CWD_{opt}^{(4)}$	Reported to be optimum for a performance measure [11]
4	$\{1, 1.2796, 1.5975, 1.8654\}$	$\{1, 0.2796, 0.3179, 0.2679\}$	$CWD_{opt*}^{(4)}$	Reported to be optimum with $\delta_1 = 1$ for a performance measure [11]
n	$\{\delta_i \mid \delta_i = \sqrt{i}$ for $1 \leq i \leq n\}$	$\{\gamma_i \mid \gamma_i = \sqrt{i} - \sqrt{i-1}$ for $2 \leq i \leq n\}$, and $\gamma_1 = 1$	Euclidean CWD in n-D or $CWD_{eu}^{(n)}$	Optimum for a performance measure [2] with scaling $\rho = \dfrac{2}{1 + \sqrt{\sum\limits_{i=1}^{n} \left(\sqrt{i} - \sqrt{i-1}\right)^2}}$

Some of the weighted distance functions, which are reported in the literature for their good proximity to Euclidean distances, are illustrated in Table 2.3.

2.3.1 Chamfering Distances in a Larger Neighborhood

In the above section, though we define CWDs with weight assignments restricted to m-neighbors, $1 \leq m \leq n$ in n-D, the distance can be generalized with weight assignments to directional movements in any arbitrary large neighborhood. We define a generalized *chamfer mask* as follows [79].

Definition 2.5 A *chamfer mask* is denoted by $\mathcal{M} = \{(\overrightarrow{v_i}, w_i) | i = 1, 2, \ldots, M\}$, where $\overrightarrow{v_i}$ is a vector formed from the origin $\bar{0}$ to a point $\bar{v_i}$ in $\mathcal{Z}^n$ and w_i is a nonnegative weight associated with $\overrightarrow{v_i}$. Further, $\bar{v_i}$ is visible from the origin $\bar{0}$, and if $\overrightarrow{v_i} \in \mathcal{M}$, then all its 2^n-symmetry vectors (defined by the points in $\phi(\bar{v_i})$ as in Definition 4.5) are also in $\mathcal{M}$ and at least one of w_i's is nonzero. A point $\bar{q}$ in a set is visible from $\bar{p}$, if there does not exist any other point on the set which lies on the straight line segment between q and p.

By definition, a chamfer mask is *central symmetric*, and using this property the definition of a mask is also trivially extended in $\mathcal{R}^n$.

Definition 2.6 As $\mathcal{M}$ is central symmetric, it is sufficient to describe it by the points ($\bar{v_i}$'s) only on the positive hyperoctant $\mathcal{G} \subset \mathcal{Z}^n$, such that $\forall \bar{u}(= (u_1, u_2, \ldots, u_n)) \in \mathcal{G}$, we have the following ordering of coordinates:

$$u_1 \geq u_2 \geq \cdots \geq u_{n-1} \geq u_n \geq 0$$

The above subset of $\mathcal{M}$ is called its *generator* and denoted by $\mathcal{M}_g$.

Example 2.1 $CWD_{eu}^{(2)}$ has the chamfer mask with the following generator in $\mathcal{R}^2$.

$$\mathcal{M}_g = \{((1, 0), 1), ((1, 1), \sqrt{2})\}$$

As shown in the above example, a distance $CWD^{(n)}(\cdot; \Delta = \{\delta_i | 1 \leq i \leq n\})$ discussed in Sect. 2.3, is defined from a chamfer mask whose $\mathcal{M}_g$ contains n points, each corresponding to a strict *type-i* neighbor in $\mathcal{G}$, with the weight δ_i.

Using a chamfer mask a path from a point $\bar{u}$ to $\bar{v}$ can be constructed by chaining the displacement vectors of the mask. Thus a distance function is defined as follows:

Definition 2.7 A *chamfer mask induced distance* (CMID) from a point $\bar{u}$ to $\bar{v}$ defined by a chamfer mask $\mathcal{M} = \{(\overrightarrow{v_i}, w_i) | i = 1, 2, \ldots, M\}$ is given by

$$d_{\mathcal{M}}^{(n)}(\bar{u}, \bar{v}) = \min \left\{ \sum_{i=1}^{M} \lambda_i w_i | \bar{v} = \bar{u} + \sum_{i=1}^{M} \lambda_i \overrightarrow{v_i} \right\} \tag{2.15}$$

By definition a CMID, $d_{\mathcal{M}}^{(n)}$, is a metric, if the distance function is total. But it may not have the property of translation invariance, and thus it is not always a norm. Its condition for being a norm gets satisfied if its hyperspheres are found to be convex. This property is elucidated by introducing the concept of an *equivalent rational ball* of a chamfer mask [61, 79].

Definition 2.8 The *equivalent rational mask* (ERM) $\mathcal{M}_Q$ of a chamfer mask $\mathcal{M} = \{(\overrightarrow{v_i}, w_i) | i = 1, 2, \ldots, M\}$ is defined as $\mathcal{M}_Q = \{(\overrightarrow{\hat{v_i}}, w_i) | i = 1, 2, \ldots, M\}$, where $\hat{v_i} = \frac{\bar{v_i}}{w_i}$.

In $\mathscr{Z}^n$, with integral nonnegative weights $\frac{\bar{v}_i}{w_i}$ is a point in $\mathscr{Q}^n$, where $\mathscr{Q}$ is the set of rational numbers. Thus, $\mathscr{M}_Q$ is named. However, the above concept is also applicable for chamfer masks with w_i's as nonnegative real numbers. The set of points of ERM being triangulated provides us a hyperpolyhedron called the *equivalent rational ball* (ERB).

Definition 2.9 The hyperpolyhedron formed by an ERM, $\mathscr{M}_Q$, is called the *equivalent rational ball* (ERB) of the chamfer mask $\mathscr{M}$ and is denoted by $\mathscr{B}_{\mathscr{M}}$.

Using above concepts, we state the condition for a CMID being a norm in the following theorem [61, 79].

Theorem 2.4 *A CMID $d_{\mathscr{M}}^{(n)}$ with the chamfer mask $\mathscr{M}$ is a norm if and only if the equivalent rational ball, $\mathscr{B}_{\mathscr{M}}$, is convex.*

When a CMID is a norm, its ERB $\mathscr{B}_{\mathscr{M}}$ is also a hypersphere of radius 1. Though we have a closed form expression of CWD (refer to Eq. (2.4)), which is a special case of CMID, a more general expression of a CMID is not yet reported. There are a few functional forms reported in 2-D and 3-D [8, 76], as discussed below.

2.3.1.1 2-D

In 2-D, Borgefors [8], provided a functional form of a distance function induced by a chamfer mask as shown in Fig. 2.1. As the mask is central symmetric, it is sufficient to describe it by only three parameters, namely a, b, and c, which are nonnegative values. We may note, only visible points are assigned a weight in the mask and its generator is given by as $\mathscr{M}_g = \{((1, 0), a), ((1, 1), b), ((2, 1), c)\}$. Let us denote this mask by $\mathscr{M}_{55}(a, b, c)$. A 5×5 chamfer mask induces a norm if certain conditions are satisfied among the parameters a, b, and c. This is stated in the following theorem [8]:

Theorem 2.5 *A 5×5 chamfer mask with the generator $\mathscr{M}_g = \{((1, 0), a),$ $((1, 1), b), ((2, 1), c)\}$ induces a norm if and only if the following conditions are satisfied:*

Fig. 2.1 A 5×5 chamfer mask in 2D

	c		c	
c	b	a	b	c
	a	0	a	
c	b	a	b	c
	c		c	

$$c \geq 2a$$
$$2c \geq 3b \tag{2.16}$$
$$c \leq a + b$$

Corresponding distance function is given by the following expression [8]:

$$d^{(2)}_{\mathcal{M}_{55}(a,b,c)}(\bar{u}) = \begin{cases} au_{(1)} + (c - 2a)u_{(2)} & \text{for } 0 \leq u_{(2)} \leq \frac{u_{(1)}}{2} \\ (c - b)u_{(1)} + (2b - c)u_{(2)} & \text{for } \frac{u_{(1)}}{2} \leq u_{(2)} \leq u_{(1)} \end{cases} \tag{2.17}$$

where $u_{(i)}, i = 1, 2$ is the ith maximum coordinate value of $\bar{u}$.

2.3.1.2 3-D

In 3-D, a $5 \times 5 \times 5$ chamfer mask can be defined by six parameters (Fig. 2.2). Its generator mask $\mathcal{M}_{555}(a, b, c, d, e, f)$ is as follows:

$$\{((1, 0, 0), a), ((1, 1, 0), b), ((1, 1, 1), c), ((2, 1, 0), d), ((2, 1, 1), e), ((2, 2, 1), f)\}$$

A chamfer mask may also be defined with a subset of these parameters, e.g., $\mathcal{M}_{555}(a, b, c, e)$ [77]. For different sets of parameters, the conditions of the masks to induce a norm are discussed in [76]. Primarily, these conditions are established by checking the length of paths from the origin to a point formed by a sequence of movements in comparison with a path formed by a sequence of same types of movements, wherever possible. For the function to be a norm, the latter path should be always shorter or equal to the length of the other path. For example, to reach a point $(2, 1, 1)$, we can make a move to $(1, 0, 0)$ with the cost a and then along the direction $(1, 1, 1)$ with the cost c. On the other hand, we can reach $(2, 1, 1)$ by directly moving along this direction with the cost e. Hence, $e \leq a + c$. Likewise different other constraints could be obtained, and some of them may be redundant. In Table 2.4, we illustrate a few chamfer masks of size $5 \times 5 \times 5$, with a set of parameters with their conditions for being a norm and also closed form expressions from [76]. Note that there are other regions in the parameter space for these masks, which may satisfy conditions of a norm. The details are available in [76].

	Z=-2				Z=-1					Z=0					Z=1					Z=2		
	f		f		f	e	d	e	f		d		d		f	e	d	e	f		f		f	
f	e	d	e	f	e	c	b	c	e	d	b	a	b	d	e	c	b	c	e	f	e	d	e	f
	d		d		d	b	a	b	d	a		0		a	d	b	a	b	d		d		d	
f	e	d	e	f	e	c	b	c	e	d	b	a	b	d	e	c	b	c	e	f	e	d	e	f
	f		f		f	e	d	e	f		d		d		f	e	d	e	f		f		f	

Fig. 2.2 A $5 \times 5 \times 5$ chamfer mask in 3D

Table 2.4 Norms of CMID with a chamfer mask of size $5 \times 5 \times 5$ in 3D

Set of parameters	Norm conditions	Distance function $d_{\mathscr{M}}(\bar{u})$	Good examples
$\{a, b, c, e\}$	$\begin{array}{ll} b \leq 2a & a+b \leq e \\ e \leq 2b & e \leq a+c \\ b+2c \leq 2e \end{array}$	$\begin{cases} au_{(1)} + (b-a)u_{(2)} + (e-b-a)u_{(3)} & \text{for } u_{(1)} \geq u_{(2)} + u_{(3)} \\ (e-c)u_{(1)} + (b+c-e)u_{(2)} + (c-b)u_{(3)} & \text{for } u_{(1)} \leq u_{(2)} + u_{(3)} \end{cases}$	$\begin{array}{cccc} a & b & c & e \\ 3 & 4 & 5 & 7 \\ 11 & 15 & 19 & 27 \\ 14 & 19 & 24 & 34 \\ 17 & 23 & 29 & 41 \end{array}$
$\{a, b, c, d, e\}$	$\begin{array}{ll} b \leq 2a & d \leq a+b \\ a+b \leq e & b+c \leq 2e \\ 2a+e \leq 2d & b+e \leq c+d \end{array}$	$\begin{cases} au_{(1)} + (d-2a)u_{(2)} + (e-d)u_{(3)} & \text{for } \frac{u_{(1)}}{2} \geq u_{(2)} \geq 0 \\ (d-b)u_{(1)} + (2b-d)u_{(2)} + (e-d)u_{(3)} & \text{for } \frac{u_{(1)}}{2} \leq u_{(2)} \leq u_{(1)} - u_{(3)} \\ (e-c)u_{(1)} + (c+b-e)u_{(2)} + (c-b)u_{(3)} & \text{for } u_{(1)} \geq u_{(2)} \geq u_{(1)} - u_{(3)} \end{cases}$	$\begin{array}{ccccc} a & b & c & d & e \\ 7 & 10 & 12 & 16 & 17 \\ 8 & 11 & 14 & 18 & 20 \\ 10 & 14 & 17 & 22 & 24 \\ 13 & 18 & 22 & 29 & 31 \end{array}$
$\{a, b, c, d, e, f\}$	$\begin{array}{ll} b \leq 2a & d \leq a+b \\ a+b \leq e & 2a+e \leq d \\ f \leq b+c & 2b+e \leq d+f \end{array}$	$\begin{cases} au_{(1)} + (d-2a)u_{(2)} + (e-d)u_{(3)} & \text{for } \frac{u_{(1)}}{2} \geq u_{(2)} \geq 0 \\ (d-b)u_{(1)} + (2b-d)u_{(2)} + (e-d)u_{(3)} & \text{for } \frac{u_{(1)}}{2} \leq u_{(2)} \leq u_{(1)} - u_{(3)}, \\ (b+e-f)u_{(1)} + (f-e)u_{(2)} + (f-2b)u_{(3)} & \text{for } u_{(1)} \geq u_{(2)} \geq u_{(1)} - u_{(3)}, \\ & u_{(3)} \leq \frac{u_{(1)}}{2} \\ (e-c)u_{(1)} + (f-e)u_{(2)} + (2c-f)u_{(3)} & \text{for } u_{(1)} \geq u_{(2)} \geq u_{(1)} - u_{(3)}, \\ & u_{(3)} \geq \frac{u_{(1)}}{2} \end{cases}$	$\begin{array}{cccccc} a & b & c & d & e & f \\ 7 & 10 & 12 & 16 & 17 & 21 \\ 10 & 14 & 17 & 22 & 24 & 30 \\ 13 & 18 & 22 & 29 & 31 & 38 \\ 23 & 32 & 39 & 51 & 55 & 68 \end{array}$

2.4 Hyperoctagonal Distances

Rosenfeld and Pfaltz in [65] proposed an interesting distance function in 2-D ($\mathscr{Z}^2$), which yields to disks of octagonal shapes. The distance function is defined as follows:

Definition 2.10 In $\mathscr{Z}^2$, paths corresponding to the *octagonal distance* $d_{oct}(\cdot)$, are constructed by the alternate neighborhood definitions namely using 4-neighbor and 8-neighbor one after another (the sequence starts from 4-neighbor). The shortest length of a path between two points formed by this alternate sequence of neighbors provides the distance value.

It is shown [65], that the distance function can be expressed in the following form and it satisfies the conditions of a metric.

$$d_{oct}(\bar{u}, \bar{v}) = \max\left(\left\lceil \frac{2}{3} \cdot d_1^{(2)}(\bar{u}, \bar{v}) \right\rceil, d_2^{(2)}(\bar{u}, \bar{v})\right)$$

The name *octagonal distance* is given as the shape of a disk of this distance function is octagonal (Fig. 2.3c). In the figure we have also shown disks of $d_1^{(2)}$ (4-neighbor distance) and $d_2^{(2)}$ (8-neighbor distance), both of which are quadrilaterals.

2.4.1 Generalized Octagonal Distances

In a generalization to the octagonal distance, the generalized octagonal distances $d(B)$ is defined [28], where the type of neighborhood at each step from points $\bar{u}$ to $\bar{v}$ in 2-D is determined by a sequence of neighborhoods $B = \{b(1), b(2), \ldots, b(n)\}, \forall i$ $b(i) = 1$ or 2, denoting either *type*-1 or *type*-2 neighborhoods in $\mathscr{Z}^2$. As it is impractical to work with an infinite sequence, we consider cyclic neighborhood sequences with the cycle length $p = |B|$ as $B = \{b(1), b(2), \ldots, b(p), b(1), b(2), \ldots, b(p), \ldots\}$. It is sufficient to denote the sequence by its first cycle, i.e., $B = \{b(1), b(2), \ldots, b(p)\}$. It may be noted that d_{oct} is a special case of $d(B)$ with

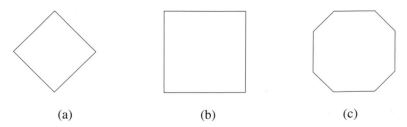

(a)   (b)   (c)

Fig. 2.3 Disks of 2-D distance functions: **a** $d_1^{(2)}$ (4-neighborhood), **b** $d_2^{(2)}$ (8-neighborhood), and **c** the octagonal distance d_{oct} (alternate neighborhoods of 4-neighbor and 8-neighbor)

$B = \{1, 2\}$ defined by under the name d_{oct}. Under this notation, m-neighbor distances in 2-D, such as $d_1^{(2)}$ and $d_2^{(2)}$ are defined by $B = \{1\}$, and $B = \{2\}$, respectively.

The metricity condition for a neighborhood sequence in 2-D is given by the following theorem [52].

Theorem 2.6 *A distance function defined by a neighborhood sequence B in $\mathscr{L}^2$ is a metric, if it satisfies the following:*
For any integer k, the sum of the first k values is not greater than the sum of k consecutive values anywhere in the sequence B.

From the above theorem, we get different metrics (or distance functions) of neighborhood length up to 4 in 2-D with neighborhood sequences $\{1\}$, $\{1, 1, 1, 2\}$, $\{1, 1, 2\}$, $\{1, 2\}$, $\{1, 1, 2, 2\}$, $\{1, 2, 2\}$, $\{1, 2, 2, 2\}$, and $\{2\}$.

2.4.2 Generalization to Hyperoctagonal Distances

Using an m-neighborhood definition, in [23, 82], the concept of generalized 2-D octagonal distances has been extended to n-D. In this case, there are n neighborhood types. Any arbitrary sequence of neighborhood types is denoted by $B = \{b(1), b(2) \ldots, b(m), \ldots\}$, where $\forall i, b(i) \in \{1, 2, \ldots, n\}$. We call the distance induced by B a *hyperoctagonal distance* (HOD). In this case also, we restrict our discussion to the cyclic neighborhood sequences[2] with a cycle length $p = |B|$ and it is represented as $B = \{b(1), b(2), \ldots, b(p)\}$.

A closed form expression for a HOD norm[3] $d_B^{(n)}(\bar{u})$ is given in [23]. The expression is quite complex and expensive for computation.[4] In the following, we state the result from [23].[5]

Theorem 2.7 *Given a neighborhood sequence $B = \{b(1), b(2), \ldots, b(p)\}$ in $\mathscr{L}^n$, the length of the minimal path from $\bar{0} \in \mathscr{L}^n$ to $\bar{u}$, $d_B^{(n)}(\bar{u})$, is given by the following expression:*

$$d_B^{(n)}(\bar{u}) = \max_{t=1}^{n} \left\{ p \left\lfloor \frac{D_t^{(n)}(\bar{u})}{f_t(p)} \right\rfloor + h(z_t; B_t) \right\} \tag{2.18}$$

where $B_t = \{b_t(1), b_t(2), \ldots, b_t(p)\}$ such that $b_t(i) = \min\{b(i), t\}$. Other terms are defined as follows:

$$f_t(j) = \begin{cases} 0 & j = 0, \\ \sum\limits_{i=1}^{j} b_t(i), & \text{for } 1 \leq j \leq p. \end{cases} \tag{2.19}$$

---

[2]In [36], the hyperoctagonal distances with non-cyclic neighborhood sequences have been studied.
[3]It is an asymptotic norm (Definition 1.3).
[4]A relatively simple expression for computing distances is provided in [53].
[5]A simple expression in 3-D was first reported in [59].

$$z_t = D_t^{(n)}(\bar{u}) \bmod f_t(p),$$ (2.20)

and,

$$h(z_t; B_t) = \min\{k \,|\, f_t(k) \geq z_t\}$$ (2.21)

The above theorem provides the distance value between any pair of points in $\mathscr{Z}^n$, as it computes the length of the shortest path. Unfortunately, not all the distance functions given a neighborhood sequence are metric. For example, in 6-dimensional spaces, for a neighborhood sequence of maximum length 6, distance functions corresponding to around 95% of all possible sequences, are found to be nonmetric. Sufficient and necessary conditions of metricity are discussed in [82]. Let us denote the set of cyclic permutation of the neighborhood sequence B by $CPERM(B)$. Then the necessary and sufficient conditions for a distance function defined by a neighborhood sequence B to become a metric are given by the following theorem [82]:

Theorem 2.8 $d_B^{(n)}$ *is a metric in* $\mathscr{Z}^n$, *if and only if* $\forall \bar{u} \in \mathscr{Z}^n$, $d_B^{(n)}(\bar{u}) \geq d_{B'}^{(n)}(\bar{u})$, $\forall B' \in CPERM(B)$.

In [23], the above conditions are further elaborated and the class of hyperoctagonal distance functions satisfying these conditions is characterized as *well-behaved* distances. However, the computation to check these conditions is quite expensive [22]. There is a special case, for which we have simple necessary conditions of metricity. This is stated in the following theorem [22, 82]:

Theorem 2.9 *In* $\mathscr{Z}^n$ *if the sequence in* B *is sorted in ascending order, the distance function defined by the neighborhood sequence* B *is a metric.*

Proof In the sorted neighborhood any *strict* $O(m)$ movement takes place after movements of lower order neighborhood type in a period of a neighborhood sequence. This makes the number of steps for arriving at destination point the largest for the sorted neighborhood sequence B among all the sequences of $CPERM(B)$. Hence, from Theorem 2.8 it follows: □

For example, in 4-D, the distance defined by the neighborhood sequence $\{1, 2, 2, 4\}$ is a metric. However, we cannot decide about the sequence $\{1, 2, 4, 2\}$, as it is not in an ascending order.

The sufficient condition for the property stated in Theorem 2.9, holds for sequence lengths up to 4.

Theorem 2.10 *For neighborhood sequences with the cycle length not more than 4, it requires to be always sorted for being a metric in* $\mathscr{Z}^n$.

To illustrate the implication of the above theorem, let us consider the HODs in $\mathscr{Z}^3$. In this case, any arbitrary sequence of three types of neighborhood may define a distance function. For the cycle lengths not more than 3, the neighborhood sequences resulting into metric are $\{1\}, \{2\}, \{3\}, \{1, 2\}, \{1, 3\}, \{2, 3\}, \{1, 1, 2\}, \{1, 1, 3\}, \{1, 2, 2\},$ $\{1, 2, 3\}$, $\{1, 3, 3\}$ and $\{2, 3, 3\}$.

A few typical closed form expressions of different distance functions in 2-D [28] and 3-D [50], are provided in Tables 2.5 and 2.6, respectively. While denoting a distance function with neighborhood sequence B, we use the notation $d_B^{(2)}$. For example, for $B = \{1, 2\}$, it is denoted as $d_{\{1,2\}}^{(2)}$. For convenience, when we have neighborhood sequences of low dimensional integral space (of dimension < 10), we simply denote B by consecutive sequence of digits of neighborhood types. Accordingly in the table,

Table 2.5 Generalized octagonal distances in 2-D (for length of the neighborhood sequence period less or equal to 3)

Distance functions ($d(B)$)	B	Closed form expressions
$d_1^{(2)}(\bar{u}, \bar{v})$	$\{1\}$	$d_4(\bar{u}, \bar{v})$
$d_2^{(2)}(\bar{u}, \bar{v})$	$\{2\}$	$d_8(\bar{u}, \bar{v})$
$d_{12}^{(2)}(\bar{u}, \bar{v})$	$\{1, 2\}$	$\max(\lceil 2d_4(\bar{u}, \bar{v})/3\rceil, d_8(\bar{u}, \bar{v}))$
$d_{112}^{(2)}(\bar{u}, \bar{v})$	$\{1, 1, 2\}$	$\max(\lceil 3d_4(\bar{u}, \bar{v})/4\rceil, d_8(\bar{u}, \bar{v}))$
$d_{122}^{(2)}(\bar{u}, \bar{v})$	$\{1, 2, 2\}$	$\max(\lceil 3d_4(\bar{u}, \bar{v})/5\rceil, d_8(\bar{u}, \bar{v}))$

Table 2.6 Hyperoctagonal distances in 3-D (for length of the neighborhood sequence period less or equal to 3)

$d_B^{(n)}$	B	Closed form expressions
$d_1^{(3)}(\bar{u}, \bar{v})$	$\{1\}$	$d_6(\bar{u}, \bar{v})$
$d_2^{(3)}(\bar{u}, \bar{v})$	$\{2\}$	$d_{18}(\bar{u}, \bar{v})$
$d_3^{(3)}(\bar{u}, \bar{v})$	$\{3\}$	$d_{26}(\bar{u}, \bar{v})$
$d_{12}^{(3)}(\bar{u}, \bar{v})$	$\{1, 2\}$	$\max(\lfloor (2d_6(\bar{u}, \bar{v}) + 2)/3\rfloor, d_{26}(\bar{u}, \bar{v}))$
$d_{13}^{(3)}(\bar{u}, \bar{v})$	$\{1, 3\}$	$\max(\lfloor (d_6(\bar{u}, \bar{v}) + 1)/4\rfloor + \lfloor (d_6(\bar{u}, \bar{v}) + 2)/4\rfloor,$ $\lfloor (2D_2^{(3)}(\bar{u}, \bar{v}) + 2)/3\rfloor, d_{26}(\bar{u}, \bar{v}))$
$d_{23}^{(3)}(\bar{u}, \bar{v})$	$\{2, 3\}$	$\max(\lfloor (2d_6(\bar{u}, \bar{v}) + 4)/5\rfloor, d_{26}(\bar{u}, \bar{v}))$
$d_{112}^{(3)}(\bar{u}, \bar{v})$	$\{1, 1, 2\}$	$\max(\lfloor (3d_6(\bar{u}, \bar{v}) + 3)/4\rfloor, d_{26}(\bar{u}, \bar{v}))$
$d_{113}^{(3)}(\bar{u}, \bar{v})$	$\{1, 1, 3\}$	$\max(\lfloor (d_6(\bar{u}, \bar{v}) + 4)/5\rfloor + \lfloor (d_6(\bar{u}, \bar{v}) + 3)/5\rfloor + \lfloor (d_6(\bar{u}, \bar{v}) + 2)/5\rfloor, \lfloor (3D_2^{(3)}(\bar{u}, \bar{v}) + 3)/4\rfloor, d_{26}(\bar{u}, \bar{v}))$
$d_{122}^{(3)}(\bar{u}, \bar{v})$	$\{1, 2, 2\}$	$\max(\lfloor (3d_6(\bar{u}, \bar{v}) + 4)/5\rfloor, d_{26}(\bar{u}, \bar{v}))$
$d_{123}^{(3)}(\bar{u}, \bar{v})$	$\{1, 2, 3\}$	$\max(\lfloor (d_6(\bar{u}, \bar{v}) + 2)/3\rfloor + \lfloor (d_6(\bar{u}, \bar{v}) + 4)/6\rfloor, \lfloor (3D_2^{(3)}(\bar{u}, \bar{v}) + 4)/5\rfloor, d_{26}(\bar{u}, \bar{v}))$
$d_{133}^{(3)}(\bar{u}, \bar{v})$	$\{1, 3, 3\}$	$\max(\lfloor (d_6(\bar{u}, \bar{v}) + 6)/7\rfloor + \lfloor (2d_6(\bar{u}, \bar{v}) + 4)/7\rfloor, \lfloor (3D_2^{(3)}(\bar{u}, \bar{v}) + 4)/5\rfloor, d_{26}(\bar{u}, \bar{v}))$
$d_{223}^{(3)}(\bar{u}, \bar{v})$	$\{2, 2, 3\}$	$\max(\lfloor (d_6(\bar{u}, \bar{v}) + 6)/7\rfloor + \lfloor (d_6(\bar{u}, \bar{v}) + 4)/7\rfloor + \lfloor (d_6(\bar{u}, \bar{v}) + 2)/7\rfloor, d_{26}(\bar{u}, \bar{v}))$
$d_{233}^{(3)}(\bar{u}, \bar{v})$	$\{2, 3, 3\}$	$\max(\lfloor (d_6(\bar{u}, \bar{v}) + 7)/8\rfloor + \lfloor (d_6(\bar{u}, \bar{v}) + 5)/8\rfloor + \lfloor (d_6(\bar{u}, \bar{v}) + 2)/8\rfloor, d_{26}(\bar{u}, \bar{v}))$

In the above, $D_2^{(3)}(\bar{u}, \bar{v}) = \max(|u_1 - v_1| + |u_2 - v_2|, |u_2 - v_2| + |u_3 - v_3|, |u_3 - v_3| + |u_1 - v_1|)$

we use $d_{12}^{(2)}$ to denote the same distance function. For historical reason, the same distance is also denoted as d_{oct}. We note that in our notation $d_m^{(n)}$ may be interpreted as an m-neighbor distance function in n-D, as well as, the distance function defined by the neighborhood sequence $B = \{m\}$ in n-D. We may observe that the distance functions given in Tables 2.5 and 2.6, are not always strictly a norm. But as the distance values tend to infinity, they asymptotically behave like a norm (Definition 1.3). Considering this asymptotic behavior we treat HODs as norms. In Sect. 2.5.3, we discuss approximations of these distance functions in the form of weighted t-cost distances (WtD), which are norms.

2.4.3 Simple Octagonal Distances in 2-D

In [20], a simpler form of expression of the distance function as described in Theorem 2.7, in $\mathscr{Z}^2$ has been provided. This is stated below.

Theorem 2.11 *Given a neighborhood sequence $B = \{b(1), b(2), \ldots, b(p)\}$ in 2-D, the length of the minimal path from $(0, 0)$ to $\bar{u} = (u_1, u_2)$ in $\mathscr{Z}^2$, $d_B^{(2)}(\bar{u})$, is given by the following expression:*

$$d_B^{(2)}(\bar{u}) = \max\{|u_1|, |u_2|, \sum_{j=1}^{p} \lfloor \frac{|u_1| + |u_2| + g(j)}{f(p)} \rfloor\} \tag{2.22}$$

where $f(0) = 0$, $f(i) = \sum_{k=1}^{i} b(k)$, $1 \le i \le p$, and $g(i) = f(p) - f(i-1) - 1$, $1 \le i \le p$.

As discussed, the above theorem provides the distance between two points in $\mathscr{Z}^2$. The necessary and sufficient conditions for the above function to be a metric are given by the following theorem [20]:

Theorem 2.12 $d_B^{(2)}(\bar{u})$ *of Theorem 2.11 is a metric iff $B = \{b(1), b(2), \ldots, b(p)\}$ is well-behaved by satisfying the following conditions:*

$$\begin{aligned} f(i) + f(j) &\le f(i+j) & i+j \le p \\ &\le f(p) + f(i+j-p) & i+j \ge p \end{aligned} \tag{2.23}$$

There are a few neighborhood sequences for which the summation part of Eq. (2.22), can be simplified by a simple ceiling ('$\lceil . \rceil$') operation . For example, for $B = \{1, 2\}$, the expression $\lfloor \frac{|u_1| + |u_2| + 1}{3} \rfloor + \lfloor \frac{|u_1| + |u_2| + 2}{3} \rfloor$ could be simplified by $\lceil \frac{2(|u_1| + |u_2|)}{3} \rceil$. Distance functions in this form are called *simple octagonal distances*. It has been shown in [20], that there exists a unique neighborhood sequence B for every length p of a sequence, and $f(p)$, $p \le f(p) \le 2p$. This is stated in the following theorem:

Theorem 2.13 $d_B^{(2)}(\bar{u})$ *is simple with the functional form* $\max\{|u_1|, |u_2|, \lceil\frac{(|u_1|+|u_2|)}{m}\rceil\}$, *iff*

$$b(k) = \lfloor\frac{kf(p)}{p}\rfloor - \lfloor\frac{(k-1)f(p)}{p}\rfloor, \text{ for } 1 \leq k \leq p \qquad (2.24)$$

where $m = \frac{f(p)}{p}$ *is a value in* $[1, 2]$, *such that* $f(p)$ *and* p *are relatively prime. Additionally,* $B = \{1\}$ *corresponding to* $m = 1$, *and* $B = \{2\}$ *corresponding to* $m = 2$ *are also simple octagonal distances.*

A simple octagonal distance is uniquely characterized by a rational number $m = \frac{q}{p} \in [1, 2]$, so that q and p are coprimes. We denote it by $d_B^{(2)}(\cdot|q, p)$. It can be shown that a simple octagonal distance is a metric. A few examples of such distances are with neighborhood sequences $\{1, 2\}$ ($d_{12}^{(2)}(\bar{u}|3, 2) = \max\{|u_1|, |u_2|, \lceil\frac{2(|u_1|+|u_2|)}{3}\rceil\}$), $\{1, 1, 2\}$ ($d_{112}^{(2)}(\bar{u}|4, 3) = \max\{|u_1|, |u_2|, \lceil\frac{3(|u_1|+|u_2|)}{4}\rceil\}$), $\{1, 1, 2, 1, 2\}$ ($d_{11212}^{(2)}(\bar{u}|7, 5) = \max\{|u_1|, |u_2|, \lceil\frac{5(|u_1|+|u_2|)}{7}\rceil\}$), etc.

2.4.4 Weighted Generalized Neighborhood Sequence

The definition of $O(m)$ neighborhood can be extended to include any arbitrary set of points around a point in $\mathcal{Z}^n$. For example, an extended neighborhood set $\mathbb{N}_K = \{(\pm 2, \pm 1), (\pm 1, \pm 2)\}$ in $\mathcal{Z}^2$ defines *Knight's move* of the game of chess. With this definition of the neighborhood, given a point $\bar{u}$, its immediate neighboring points are $\{(u_1 \pm 2, u_2 \pm 1), (u_1 \pm 1, u_2 \pm 2)\}$. In [26], it was shown that the distance defined by $\mathbb{N}_K$ is a metric. The corresponding distance function is given by the following equation [26]:

$$d_{knight}(\bar{u}) = \begin{cases} \max\left(\lceil\frac{u_{(1)}}{2}\rceil, \lceil\frac{(u_{(1)}+u_{(2)})}{3}\rceil\right) + (u_{(1)} + u_{(2)}) \\ -\max\left(\lceil\frac{u_{(1)}}{2}\rceil, \lceil\frac{(u_{(1)}+u_{(2)})}{3}\rceil\right) \bmod 2 & \bar{u} \notin \{(\pm 1, 0), (0, \pm 1), \\ & (\pm 2, \pm 2)\} \\ 3 & \bar{u} \in \{(\pm 1, 0), (0, \pm 1)\} \\ 4 & \bar{u} \in \{(\pm 2, \pm 2)\} \end{cases}$$
$$(2.25)$$

where $u_{(1)} = \max(|u_1|, |u_2|)$ and $u_{(2)} = \min(|u_1|, |u_2|)$.

The definition of a generalized neighborhood of a point $\bar{u} \in \mathcal{Z}^n$ is given as follows:

Definition 2.11 A *generalized neighborhood* $\mathbb{N}$, a subset of $\mathcal{Z}^n$, defines $\bar{v} \in \mathcal{Z}^n$ a neighbor of $\bar{u} \in \mathcal{Z}^n$, if $\exists \bar{p} \in \mathbb{N}$, such that $\bar{v} = \bar{u} + \bar{p}$. $\mathbb{N}$ maintains 2^n-Symmetry (refer to Definition 4.5), so that $\bar{p} \in \mathbb{N} \implies \phi(\bar{p}) \subset \mathbb{N}$.

A generalized neighborhood definition may not ensure reachability of every point in the space from a point. For example, given $\mathbb{N} = \{(\pm 1, \pm 1)\}$ in $\mathcal{Z}^2$, we will not be able to reach at $(0, 1)$ from $(0, 0)$. The move corresponds to the Bishop's move in the game of chess. Hence, not every $\mathbb{N}$ defines a distance function. Following theorem states the necessary and sufficient conditions for a generalized neighborhood to define a metric.[6]

Theorem 2.14 *A generalized neighborhood $\mathbb{N}$ in $\mathscr{Z}^n$, defines a distance function, which is a metric, if and only if, there exists a finite path following this neighborhood definition from the origin $(\bar{0})$ to a point in its $O(1)$ neighborhood.*

The above theorem explains why the generalized neighborhood corresponding to a Knight's move in chess defines a metric, but those for a Bishop do not satisfy the conditions. A few more interesting results in 2-D are reported in [30]. Next, we define a weighted generalized neighborhood following the definition given in [83].

Definition 2.12 A *weighted generalized neighborhood* $\mathbb{P}$ is defined as $(\mathbb{N}, W(\cdot))$, by associating a weight function $W : \mathbb{N} \to \mathscr{R}^+$ to $\mathbb{N}$. The association of weights is also centrally symmetric so that for any point $\bar{p} \in \mathbb{N}$, $W(\bar{p}) \implies \forall \bar{q} \in \phi(\bar{p})$, $W(\bar{q}) = W(\bar{p})$.

From the above definitions, we easily observe that $O(m)$ neighbors in $\mathscr{Z}^n$ are subsets of $\{0, \pm 1\}^n$, and the weights are uniformly associated to the neighboring points with the value 1. $\mathbb{P}$ also defines a chamfer mask (Definition 2.5).

Definition 2.13 The *equivalent chamfer mask (ECM)* of $\mathbb{P} = (\mathbb{N}, W(\cdot))$ is given by $\mathscr{M}_{\mathbb{P}} = \{(\overrightarrow{v}, w)|\bar{v} \in \mathbb{N}, W(\bar{v}) = w\}$.

Hence, distances defined by $\mathbb{P}$ would be the same as obtained from its equivalent chamfer mask.[7]

Definition 2.14 A *weighted generalized neighborhood sequence* (WGNS) is a periodic sequence of $\{\mathbb{P}_1, \mathbb{P}_2, \ldots, \mathbb{P}_p\}$, where $\mathbb{P}_i$ defines a weighted generalized neighborhood. The distance induced by a WGNS is the length of the shortest path, if exists, between two points.

From [83], we define an ECM for a WGNS.

Definition 2.15 Given a weighted generalized neighborhood sequence (WGNS) $\mathbb{B} = \{\mathbb{P}_1, \mathbb{P}_2, \ldots, \mathbb{P}_p\}$, of the period p, its *equivalent chamfer mask* (ECM) is given by $\mathscr{M}_{\mathbb{B}} = \{(\overrightarrow{v}, w)|\bar{v}$ is reachable from $\bar{0}$ by atmost p steps, and the distance of $\bar{v}$ from $\bar{0} = w\}$.

The above definition is very useful in relating a WGNS with a CMID, in particular, relating a HOD to an equivalent CMID. Let us illustrate this relation by an example.

---

[6]A proof of this theorem in $\mathscr{Z}^2$ is given in [30]. The same argument can be extended in favor of the theorem.

[7]In this definition, the condition of the visibility of a point in the neighboring set is relaxed. But for a consistent definition, it should satisfy the condition of homogeneity along a direction.

Example 2.2 Consider the neighborhood sequence $B = \{1, 1, 2\}$ in $\mathscr{Z}^2$ with unit weight assignments to $O(m)$, $m = 1, 2$ neighboring points. Its ECM is given by a chamfering mask with the following generator.

$$\mathscr{M}_g = \{((1, 0), 1), ((2, 0), 2), ((3, 0), 3), ((1, 1), 2), ((2, 2), 3), ((2, 1), 3), ((3, 1), 3)\}$$

We may observe that the corresponding ERB of this chamfer mask is not convex, and hence strictly speaking $d^{(2)}_{112}$ is not a norm.

But if the function for weight assignment is given by $d(\bar{u}) = \max(\max(|u_1|, |u_2|), \frac{3}{4}(|u_1| + |u_2|))$, the ECM is given by the following generator:

$$\{((1, 0), 1), ((2, 0), 2), ((3, 0), 3), ((1, 1), 1.5), ((2, 2), 3), ((2, 1), 2.25), ((3, 1), 3)\}.$$

It forms a convex ERB, and hence the respective WGNS is a norm. We may note $d^{(2)}_{112}$ asymptotically behaves like a norm.

Theorem 2.15 *A WGNS* $\mathbb{B}$ *defines a metric, if and only if, the generalized neighborhood* $\mathbb{N}$ *corresponding to its ECM* $\mathscr{M}_{\mathbb{B}}$ *satisfies the criteria of Theorem 2.14. Further, if the ERB of the ECM is convex, it is also a norm.*

Special cases of WGNS in 2-D are discussed in a few relatively recent works [55, 57, 69, 70].

2.5 Weighted t-Cost Distance Function

In [42], the class of t-cost distances is further generalized to a class of *weighted t-cost distance* (WtD) functions. Its definition is given below.

Definition 2.16 A weighted t-cost norm $WtD^{(n)}(\bar{u}; W)$ in $\mathscr{Z}^n$ is defined as

$$WtD^{(n)}(\bar{u}; W) = \max_{1 \leq t \leq n} \left\{ w_t . D_t^{(n)}(\bar{u}) \right\} \tag{2.26}$$

W is a finite set of n weights $\{w_1, w_2, \ldots, w_n\}$, where w_t's are nonnegative real numbers.

It may be noted that distance values computed by $WtD^{(n)}(\bar{u}; W)$ are nonnegative real numbers. The modification to its form for integral distance values could be made in various ways. One such form as used in [42], is given below

$$\widetilde{WtD^{(n)}}(\bar{u}; W) = \max_{1 \leq t \leq n} \left\{ \left\lceil w_t . D_t^{(n)}(\bar{u}) \right\rceil \right\} \tag{2.27}$$

The metricity of $WtD^{(n)}(\bar{u}; W)$ is presented in the following theorem:

Theorem 2.16 $WtD^{(n)}(\bar{u}; W)$ *is a metric in* $\mathscr{Z}^n$.

Proof As $D_t^{(n)}(\bar{u})$ is a metric in $\mathscr{Z}^n$,

$$\forall \bar{u} \in \mathscr{Z}^n, WtD^{(n)}(\bar{u}; W) \geq 0 \text{ and } WtD^{(n)}(\bar{u}; W) = 0, \text{ iff } u_i = 0, \forall i. \quad (2.28)$$

The condition for triangular inequalities could be proved as follows:

$$WtD^{(n)}(\bar{u}; W) + WtD^{(n)}(\bar{v}; W) = \max_{1 \leq t \leq n} \left\{ w_t D_t^{(n)}(\bar{u}; W) \right\} + \max_{1 \leq t \leq n} \left\{ w_t D_t^{(n)}(\bar{v}; W) \right\}$$

Let

$$w_{imax} D_{imax}^{(n)}(\bar{u}) = \max_{1 \leq t \leq n} \left\{ w_t . D_t^{(n)}(\bar{u}; W) \right\}, imax \in \{1, 2, \ldots, n\}$$

and,

$$w_{jmax} D_{jmax}^{(n)}(\bar{v}) = \max_{1 \leq t \leq n} \left\{ w_t . D_t^{(n)}(\bar{v}; W) \right\}, jmax \in \{1, 2, \ldots, n\}.$$

Hence,

$$WtD^{(n)}(\bar{u}; W) = w_{imax} D_{imax}^{(n)}(\bar{u}), \text{ and}$$

$$WtD^{(n)}(\bar{v}; W) = w_{jmax} D_{jmax}^{(n)}(\bar{v}).$$

Again,

$$WtD^{(n)}(\bar{u} + \bar{v}; W) = \max_{1 \leq t \leq n} \left\{ w_t D_t^{(n)}(\bar{u} + \bar{v}) \right\}$$

Let

$$WtD^{(n)}(\bar{u} + \bar{v}; W) = w_{kmax} D_{kmax}^{(n)}(\bar{u} + \bar{v})$$

As $D_{kmax}^{(n)}(\bar{u})$ is a metric,

$$w_{kmax} D_{kmax}^{(n)}(\bar{u}) + w_{kmax} D_{kmax}^{(n)}(\bar{v}) \geq w_{kmax} D_{kmax}^{n}(\bar{u} + \bar{v})$$

Again, from the definitions of weighted t-cost distances,

$$w_{imax} D_{imax}^{(n)}(\bar{u}) \geq w_{kmax} D_{kmax}^{(n)}(\bar{u})$$

and

$$w_{jmax} D_{jmax}^{(n)}(\bar{v}) \geq w_{kmax} D_{kmax}^{(n)}(\bar{v}).$$

Hence,

$$w_{imax} D_{imax}^{(n)}(\bar{u}) + w_{jmax} D_{jmax}^{(n)}(\bar{v}) \geq w_{kmax} D_{kmax}^{(n)}(\bar{u} + \bar{v}).$$

Hence, the theorem.   □

In the same way, $\widetilde{WtD}^{(n)}(\bar{u}; W)$ is found to be a norm. As there is a very little constraint in the elements of W, which should be nonnegative, and at least one of them is nonzero, WtDs define a large class of distance functions. A suitable assignment of weights may lead to a good approximation of Euclidean metric.

2.5.1 A Few Special Cases

There exist a few examples of WtD, which have good approximation properties of Euclidean norms. For example, by minimizing the upper bound of *maximum relative error* (MRE) of WtD (discussed in the next chapter) we obtain the following weight assignment.

$$w_t = \begin{cases} \frac{2}{\sqrt{t}+\frac{t}{\sqrt{n}}} & t \leq \sqrt{n}. \\ \frac{2}{1+\sqrt{t}} & \sqrt{n} \leq t \leq n. \end{cases} \tag{2.29}$$

We refer to the above WtD as *simple upper bound optimized* weighted t-cost norm, and denote it as $WtD_{sub}^{(n)}$. There exists another interesting weight assignment, which gives much lower strict upper bound of MRE. In this case, the weights are assigned as given in Eq. (2.30).

$$w_t = \frac{1}{\sqrt{t}}, \text{ for } 1 \leq t \leq n. \tag{2.30}$$

In Sect. 4.5.2, it is shown that with this set of weights, the WtD norm exhibits good proximity with the Euclidean norm. We call this norm *inverse square root weighted t-cost norm* and denote it as $WtD_{isr}^{(n)}$. Its notation and definition are given below

$$WtD_{isr}^{(n)}(\bar{u}) = WtD^{(n)}\left(\bar{u}; W_{isr} = \left\{w_t | w_t = \frac{1}{\sqrt{t}}\right\}\right) = \max_{1 \leq t \leq n}\left\{\frac{D_t^{(n)}(\bar{u})}{\sqrt{t}}\right\} \tag{2.31}$$

In the same way, *integer inverse square root weighted t-cost norm* is defined as

$$\widetilde{WtD}_{isr}^{(n)}(\bar{u}) = \max_{1 \leq t \leq n}\left\{\left\lceil \frac{D_t^{(n)}(\bar{u})}{\sqrt{t}} \right\rceil\right\} \tag{2.32}$$

2.5.2 Other Distances as Special Cases of WtD

From the functional form of *m-neighbor norms* (Eq. (2.1)), we observe that they are also included in the class of *weighted t-cost norms*. As *m-neighbor norm* is expressed as [24]

$$d_m^{(n)}(\bar{u}) = \max\left(D_1^n(\bar{u}), \left\lceil \frac{D_n^{(n)}(\bar{u})}{m} \right\rceil\right), \text{ for } 1 \le m \le n.$$

we get an equivalent weight assignment for a WtD so that $w_1 = 1$, $w_n = \frac{1}{m}$ and all other weights are zero. Likewise, *weighted t-cost norm* becomes usual *t-cost norm*, when $w_t = 1$ and all other weights are zero. Hence, they provide a general class of metrics with infinite possibilities including both *m-neighbor* and *t-cost* norms.[8]

2.5.3 Approximation of Hyperoctagonal Distances by Weighted t-cost Norms

In Sect. 2.4.2, we discuss a closed form expression of a HOD in Eq. (2.18), given by a cyclic neighborhood sequence $B = (b(1), b(2), \ldots, b(p))$ of the length of the cycle $\mid B \mid = p$.

In [42], a simple approximation of the corresponding distance function is suggested in the form of a *weighted t-cost norm*. This is given below

$$\hat{d}_B^{(n)}(\bar{u}) = \max_{t=1}^{n}\left\{ \left\lceil \frac{p D_t^{(n)}(\bar{u})}{f_t(p)} \right\rceil \right\} \tag{2.33}$$

where

$$f_t(p) = \sum_{i=1}^{p} min(b(i), t) \tag{2.34}$$

We state the bounds of the approximations in the following theorem:

Theorem 2.17

$$0 \le |d_B^{(n)}(\bar{u}) - \hat{d}_B^{(n)}(\bar{u})| < p \tag{2.35}$$

where p is the length of B.

Proof Consider the expressions in Eqs. (2.18) and (2.33). We find that

$$0 \le \left\lceil p \frac{D_t^{(n)}(\bar{u})}{f_t(p)} \right\rceil - p \left\lfloor \frac{D_t^{(n)}(\bar{u})}{f_t(p)} \right\rfloor \le p. \tag{2.36}$$

Again, $\forall t, h(z_t; B_t) \le p$ (from Eq. (2.20)).
Hence, the theorem.   □

From the above theorem, we find that $d_B^{(n)}(\bar{u})$ and $\hat{d}_B^{(n)}(\bar{u})$ are equal when $p = 1$. This is stated in the following corollary:

---

[8]In [25] *m*-neighbor and *t*-cost distances are generalized by a class *t-cost m-neighbor norms*.

Corollary 1 *Given* $B = \{m\}$, $1 \leq m \leq n$, $d_B^{(n)}(\bar{u}) = \hat{d}_B^{(n)}(\bar{u}) = d_m^{(n)}(\bar{u})$.

Proof In [18], it is proved that

$$d_{\{m\}}^{(n)}(\bar{u}) = d_m^{(n)}(\bar{u}) = \max\left(D_1^{(n)}(\bar{u}), \left\lceil \frac{D_n^{(n)}(\bar{u})}{m} \right\rceil\right) \qquad (2.37)$$

Now, the weight vector W for $\hat{d}_{\{m\}}^{(n)}(\bar{u})$ is defined as follows:

$$w_t = \begin{cases} \frac{1}{t} & t \leq m, \\ \frac{1}{m} & m < t \leq n \end{cases} \qquad (2.38)$$

Let us consider the following identities:

$$\max_{t=1}^{m}\left\{\left\lceil \frac{D_t^{(n)}(\bar{u})}{t} \right\rceil\right\} = D_1^{(n)}(\bar{u}), \qquad (2.39)$$

and

$$\max_{t=m}^{n}\left\{\left\lceil \frac{D_t^{(n)}(\bar{u})}{m} \right\rceil\right\} = \left\lceil \frac{D_n^{(n)}(\bar{u})}{m} \right\rceil. \qquad (2.40)$$

From above two identities, we get

$$\hat{d}_{\{m\}}^{(n)}(\bar{u}) = \max\left(D_1^{(n)}(\bar{u}), \left\lceil \frac{D_n^{(n)}(\bar{u})}{m} \right\rceil\right) \qquad (2.41)$$

Hence, the corollary.   □

From the above corollary, we get another interpretation of m-neighbor distances through the definition of weighted t-cost norms.

A few examples of approximate octagonal distance functions by specifying their weight vector W are provided in Table 2.7. In [42], the proximity to the exact expression of HODs is studied by observing the *maximum absolute deviation* of the approximated value from the true value of the distance.

Definition 2.17 The *maximum absolute deviation* $(\Delta_{max}^{(n)})$ between two distance functions ρ_1 and ρ_2 in a bounded region is computed as follows:

$$\Delta_{max}^{(n)}(\rho_1, \rho_2) = \max_{0 \leq i_n \leq i_{n-1} \leq \ldots \leq i_1 \leq M} \{|\ \rho_1((i_1, i_2, \ldots, i_n)) - \rho_2((i_1, i_2, \ldots, i_n))\ |\} \qquad (2.42)$$

From [42], in Table 2.7, the maximum absolute deviation $(\Delta_{max}^{(n)}(d_B^{(n)}, \hat{d}_B^{(n)}))$ from the true distance values for different neighborhood sequences in 2-D and 3-D are shown. The value of M is taken as 512. The errors are found to be very small. It

Table 2.7 Maximum absolute deviation between $d_B^{(n)}$ and $\hat{d}_B^{(n)}$

n	B	W for $\hat{d}_B^{(n)}(.)$	$\Delta_{max}^{(n)}(d_B^{(n)}, \hat{d}_B^{(n)})$
2	{12}	$\{1, \frac{2}{3}\}$	0
	{112}	$\{1, \frac{3}{4}\}$	0
	{1112}	$\{1, \frac{4}{5}\}$	0
	{122}	$\{1, \frac{3}{5}\}$	0
3	{1222}	$\{1, \frac{4}{7}\}$	0
	{12}	$\{1, \frac{2}{3}, \frac{2}{3}\}$	0
	{23}	$\{1, \frac{1}{2}, \frac{2}{5}\}$	0
	{13}	$\{1, \frac{2}{3}, \frac{1}{2}\}$	1
	{113}	$\{1, \frac{3}{4}, \frac{3}{5}\}$	1
	{223}	$\{1, \frac{1}{2}, \frac{3}{7}\}$	0
	{11123}	$\{1, \frac{5}{7}, \frac{5}{8}\}$	1
	{1111223}	$\{1, \frac{7}{10}, \frac{7}{11}\}$	1

indicates that differences between the approximate and exact expressions for HODs are negligible in all practical purposes. In subsequent chapters, we discuss how these approximations are useful in computing errors of hyperoctagonals distances from Euclidean metrics.

2.5.4 Linear and Nonlinear Combination of Distance Functions

In [65], it is observed that a sum of a set of metrics is also a metric. This is also true for the maximum of the set. These are the two examples of a linear and a nonlinear combinations of distance functions, which preserve the metricity properties. We can trivially extend these observations in the following theorems:

Theorem 2.18 *Given a set of distance functions $\mathcal{D} = \{d_1, d_2, \ldots, d_k\}$ in $\mathscr{L}^{(n)}(\mathscr{R}^{(n)})$ and a set of nonnegative weights $W = \{w_1, w_2, \ldots, w_k\}$ such that at least one of them is nonzero, a linear combination of $\mathcal{D}$ in the form $\sum_{i=1}^{k} w_i d_i$ and a nonlinear combination in the form $\max_{i=1}^{k}\{w_i d_i\}$ are metrics.*

We observe that WtD is a nonlinear combination of t-cost distances preserving its metricity. A comprehensive discussion on metricity preserving transform is available in [50]. In Chap. 5, we discuss how a linear combination of metrics is useful in approximating Euclidean metrics. In particular, a linear combination of a WtD and a CWD is found to be a powerful representation in analyzing properties of distance

Fig. 2.4 A hierarchy of
classes of distance functions

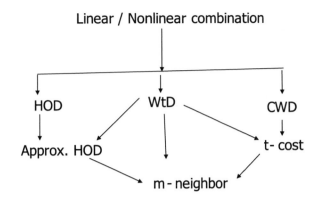

functions. It generalizes several classes of distance functions such as m-neighbor, t-cost, hyperoctagonal distances in addition to WtD and CWD. A hierarchy of these distance functions is shown in Fig. 2.4.

2.6 Concluding Remarks

In this chapter, we discuss a few representative classes of distance functions. There are a few other classes, which define distance functions extending some of the concepts discussed above. For example, in [25], a generalization of t-cost and m-neighbor distances in $\mathscr{Z}^n$ is proposed. In [36], hyperoctagonal distances of neighborhood sequence of infinite length of cycle has been explored.

There are studies on distances in a space of 2-D *non-square* [54] grids and, 3-D *non-cubic* [71], and *non-standardized* [72–75] grids. However, in this book, we restrict our discussion on distance functions defined in $\mathscr{Z}^n$ with the uniform hyper-cubic tessellation of the space. Their trivial extensions in the continuous real space $\mathscr{R}^n$ are also considered. In subsequent chapters, we discuss analysis of errors of the distance functions as reviewed in this chapter.

Chapter 3
Error Analysis: Analytical Approaches

In the last chapter, we discuss about different types of digital distance functions reported in the literature. In this chapter, we consider how these functions approximate the Euclidean distance in the same dimensional space by analyzing and estimating their differences from corresponding Euclidean distance values at a point in the space. We follow analytical approaches for studying this proximity. In some distance functions, it may not be possible to perform precise analytical treatment for deriving an expression of analytical error measures in a generic form. In such cases, we also consider empirical analysis to observe their closeness to Euclidean metrics.

3.1 Analytical Error Measures

While defining error measures, we use the norm of a distance function, as all the distance functions of our interests are either norms or asymptotic norms. There are usually two types of analytical errors, that could be defined at every point in the space, namely *absolute error* and *relative error*.

Definition 3.1 The *absolute error* of a distance function d at a point $\bar{u}$ in n-D is defined as follows:

$$a(\bar{u}) = |E^{(n)}(\bar{u}) - d(\bar{u})|$$

For a bounded region, if the coordinates are within $[0, M]$, we define *normalized absolute error* (NAE) as follows:

Definition 3.2 The *normalized absolute error* (NAE) of a distance function d at a point $\bar{u}$ in $\mathscr{Z}_{\mathscr{M}}^n$ is defined as follows:

© The Author(s), under exclusive license to Springer Nature Singapore Pte Ltd. 2020
J. Mukhopadhyay, *Approximation of Euclidean Metric by Digital Distances*,
https://doi.org/10.1007/978-981-15-9901-9_3

$$a_M(\bar{u}) = \frac{a(\bar{u})}{M}$$

Definition 3.3 The *relative error* of a distance function d at a point $\bar{u}$ in n-D is defined as follows:

$$r(\bar{u}) = \frac{a(\bar{u})}{E^{(n)}(\bar{u})}$$

The relative error at origin $(\bar{0})$ is taken as 0.

Using one of the above errors in distances, we define various aggregated form of analytical errors in $\mathscr{L}^n$ as defined in [27].

Definition 3.4 $REL_i(d(\cdot), n)$ represents the *maximum of relative error of type: i*, $0 \leq i \leq 4$ for a given distance function d in n-D

$$
\begin{aligned}
REL_0(d(\cdot), n) &= \max_{\bar{u} \in \mathscr{L}^n} r(\bar{u}) \\
REL_1(d(\cdot), n) &= \max_{\bar{u} \in \mathscr{L}^n} \frac{a(\bar{u})}{d(\bar{u})} \\
REL_2(d(\cdot), n) &= \max_{\bar{u} \in \mathscr{L}^n} \frac{a(\bar{u})}{\max\{d(\bar{u}), E^{(n)}(\bar{u})\}} \quad\quad (3.1) \\
REL_3(d(\cdot), n) &= \max_{\bar{u} \in \mathscr{L}^n} \frac{a(\bar{u})}{|d(\bar{u}) + E^{(n)}(\bar{u})|} \\
REL_4(d(\cdot), n) &= \max_{\bar{u} \in \mathscr{L}^n} \frac{a(\bar{u})}{\sqrt{d(\bar{u})^2 + E^{(n)}(\bar{u})^2}}
\end{aligned}
$$

Note that unlike $REL_{\{0,1\}}(d(\cdot), n)$, the last three relative errors $REL_{\{2,3,4\}}(d(\cdot), n)$ are normalized

$$REL_i \leq 1, \quad\quad 2 \leq i \leq 4$$

In our discussion, we mostly use $REL_0(d(\cdot), n)$, which is called *maximum relative error (MRE)*, and is restated as follows:

Definition 3.5 The *maximum relative error (MRE)* of a distance function d in $\mathscr{L}^n$ with respect to the Euclidean norm $E^{(n)}$ is the upper bound of the relative error of the norm. This is given by the following expression:

$$MRE(d) = \max_{\bar{u} \in \mathscr{L}^n} \left\{ \frac{|E^{(n)}(\bar{u}) - d(\bar{u})|}{E^{(n)}(\bar{u})} \right\} \quad\quad (3.2)$$

Similarly, *maximum normalized absolute error* (MNAE) of a distance function is defined.

Definition 3.6 The *maximum normalized absolute error* (MNAE) of a distance function d in $\mathscr{L}_M^n$ with respect to the Euclidean norm $E^{(n)}$ is the upper bound of the normalized error. This is given by the following expression:

$$MNAE(d) = \max_{\bar{u} \in \mathscr{Z}_M{}^n} \left\{ \frac{|E^{(n)}(\bar{u}) - d(\bar{u})|}{M} \right\} \tag{3.3}$$

We also use the average of relative errors in a space as defined below.

Definition 3.7 The *average relative error (ARE)* of a distance function d in $\mathscr{Z}^n$ with respect to the Euclidean norm $E^{(n)}$ is the average of the relative error of the norm. This is given by the following expression:

$$ARE(d) = \mathscr{E} \left\{ \frac{|E^{(n)}(\bar{u}) - d(\bar{u})|}{E^{(n)}(\bar{u})} \right\} \tag{3.4}$$

where $\mathscr{E}\{x\}$ denotes expectation of a random variable x.

All the above error measures, though defined in $\mathscr{Z}^n$, are also trivially extended in $\mathscr{R}^n$. We would refer to them by the same nomenclatures and interpret their meaning in the context of the geometric space.

3.2 Error Analysis of m-Neighbor Distances

For an m-neighbor distance $d_m^{(n)}$, the MRE (Definition 3.5) is bounded by a real constant [27]. But the absolute error has no upper bound. These are stated in the following theorems. The proofs are provided in [27].

Theorem 3.1 *The absolute error $a(\bar{u})$ is unbounded, that is, for all $M \in \mathscr{R}^+$, $\exists \bar{u} \in \mathscr{Z}^n$ such that $a(\bar{u}) = |E^{(n)}(\bar{u}) - d_m^{(n)}(\bar{u})| > M$.*

Theorem 3.2 $\forall m, n, \ 1 \leq m \leq n$,

$$REL_0(d_m^{(n)}, n) = \max_{\bar{u} \in \mathscr{Z}^n} r(\bar{u}) < \frac{n}{m}. \tag{3.5}$$

In our subsequent discussion, we use only the relative errors to measure the goodness of digital approximation to Euclidean distance. Following are the results derived in [27]:

Corollary 3.1 $\forall n, \forall m, \ 1 \leq m \leq n, REL_{\{1,2,3,4\}}(d_m^{(n)}, n)$ *are all bounded. The bounds are as follows:*

$$REL_1(d_m^{(n)}, n) < \max \left\{ \sqrt{m} - 1, 1 - \frac{n}{n+m} \right\}$$

$$REL_2(d_m^{(n)}, n) < \max \left\{ 1 - \frac{1}{\sqrt{m}}, 1 - \frac{n}{n+m} \right\} < 1$$

$$REL_3(d_m^{(n)}, n) < \max \left\{ \frac{\sqrt{m}-1}{\sqrt{m}+1}, \frac{n}{n+2m} \right\} < 1$$

$$REL_4(d_m^{(n)}, n) < \max \left\{ \frac{\sqrt{m}-1}{\sqrt{m}+1}, \frac{n}{\sqrt{n^2+2nm+2m^2}} \right\} < 1$$

Table 3.1 Optimal m for least relative errors. m_i^{opt} denotes optimal m for $REL_i(d_m^{(n)}, n), 0 \le i \le 4$

Dimension	1	2	3	4	5	6	7	8	9	10
m_0^{opt}	1	2	2	2	2	2	3	3	3	3
m_1^{opt}	1	1	2	2	2	2	2	2	2	2
m_2^{opt}	1	1	2	2	2	2	2	2	2	2
m_3^{opt}	1	1	2	2	2	2	2	2	2	2
m_4^{opt}	1	1	2	2	2	2	2	2	2	2

The upper bound on $REL_0(d_m^{(n)}, n)$ as obtained in Theorem 3.2, is rather loose and tends to suggest that $d_m^{(n)}$ gets better as m increases and equals n. However, it is shown in [27], that this is not true and $REL_0(d_m^{(n)}, n)$ indeed minimizes for an m between 1 and n. Based on the analysis, optimal m for minimizing these errors are reported in [27]. A few of them in low dimensional spaces are illustrated in Table 3.1.

It is observed that the optimum value of m rises very slowly with increasing dimension. For $REL_{\{1,2,3,4\}}(d_m^{(n)}, n)$ the rate of this rise is slower than that of $REL_0(d_m^{(n)}, n)$, and is the slowest for $REL_1(d_m^{(n)}, n)$. For example, at dimension 40, optimum values of m for $REL_{\{0,1,2,3,4\}}(d_m^{(n)}, n)$ are 5, 2, 4, 4, and 4, respectively [27]. In Sect. 5.6.3.1, we discuss a geometric approach for computing optimal m of MRE (i.e $REL_0(d_m^{(n)}, n)$).

3.2.1 Error Analysis of Real m-Neighbor Distances

For an m-neighbor distance, the function is a metric even if m is a real number. We denote the respective real m-neighbor distance function in an n-dimensional space by $\delta_m^{(n)}$. It is possible to extend the error analysis in this case also. In this context, it is observed that the absolute normalized error (Definition 3.6) of $\delta_m^{(n)}$ is bounded. This has been stated in the following theorem. The proof of this theorem is available in [21].

Theorem 3.3

$$
\begin{aligned}
MNAE(\delta_m^{(n)}) &= \max\left(\frac{n}{m} - \sqrt{n}, \sqrt{\lfloor m \rfloor + (m - \lfloor m \rfloor)^2} - 1\right) \\
&= \max\left(\sqrt{n}\left(\frac{r}{r_I} - 1\right), \frac{r_C}{r} - 1\right) \\
MRE(\delta_m^{(n)}) &= \max\left(\frac{\sqrt{n}}{m} - 1, 1 - \frac{1}{\sqrt{(\lfloor m \rfloor + (m - \lfloor m \rfloor)^2)}}\right) \\
&= \max\left(\frac{r}{r_I} - 1, 1 - \frac{r}{r_C}\right)
\end{aligned}
\tag{3.6}
$$

where r_I and r_C are the radii of the inscribed and circumscribed hyperspheres, respectively.

3.3 Error Analysis of t-Cost Distances

We present the results from the error analysis of $D_t^{(n)}$ with respect to $E^{(n)}$ from [31], in the following theorems. The proofs of these theorems are provided in [31].

Theorem 3.4 *Absolute error $a(\bar{u})$ is unbounded, that is, $\forall M$, $M \in \mathcal{R}^+$, $\exists \bar{u} \in \mathcal{Z}^n$ such that $a(\bar{u}) = |E^{(n)}(\bar{u}) - D_t^{(n)}(\bar{u})| > M$.*

We also find that the ratio of t-cost distance with the Euclidean distance at a point lies within an interval in positive real axis [31]. This implies that the proportional error is bounded.

Theorem 3.5 *Proportional error is bounded over n-D space, that is, $\forall n \geq 1$, $1 \leq t \leq n \ni C_1, C_2 \in \mathcal{R}^+$ such that $\forall \bar{u} \in \mathcal{Z}^n - \{0\}$*

$$C_1.D_t^{(n)}(\bar{u}) \leq E^{(n)}(\bar{u}) \leq C_2.D_t^{(n)}(\bar{u}). \tag{3.7}$$

Using further analysis, the MRE of t-cost distances is derived in [31].

Theorem 3.6 *$\forall n$, $n \geq 1$, $1 \leq t \leq n$, $MRE(D_t^{(n)}) = \max(\sqrt{t} - 1, 1 - \frac{t}{\sqrt{n}})$.*

In the next chapter, we discuss how more precise the expressions for the MRE become following a geometric approach.

3.3.1 Optimal Choice of t

For every n, there exist n different choices of the associated cost t, $1 \leq t \leq n$. For every cost t, we obtain the maximum relative error (MRE). At an optimum cost t_{opt} the MRE is minimized. That is,

$$MRE(D_{topt}^{(n)}) = \min_{t=1}^{n} MRE(D_t^{(n)}). \tag{3.8}$$

This t_{opt} clearly gives the best $D_t^{(n)}$. In the following theorem, it is shown that for all n the value of t_{opt} lies between 1 and 3 [31].

Theorem 3.7 *$\forall n$, $n \geq 1$, we have $1 \leq t_{opt} \leq 3$.*

Proof We note that, $MRE(D_t^{(n)})$ minimizes for the solution of $\sqrt{t} - 1 = 1 - \frac{t}{\sqrt{n}}$. As the right hand side is < 1, $t < 4$. $\square$

We provide a few examples of optimum t-cost distance function in two and three-dimensional spaces.

Example 3.1 For $n = 2$, $MRE(D_1^{(2)}) = 1 - \frac{1}{\sqrt{2}} = 0.2929 < 0.4142 = \sqrt{2} - 1 = MRE(D_2^{(2)})$. Thus, $D_1^{(2)}$ (or chessboard distance) is closer to the Euclidean distance function than $D_2^{(2)}$ (or city block distance) in 2-D.

For $n=3$, $MRE(D_1^{(3)})) = 1 - \frac{1}{\sqrt{3}} = 0.4226$, $MRE(D_2^{(3)}) = \sqrt{2} - 1 = 0.4142$, and $MRE(D_3^{(3)}) = \sqrt{3} - 1 = 0.7321$. Hence, $D_2^{(3)}$ is better than both $D_1^{(3)}$ or lattice distance and $D_3^{(3)}$ or grid distance in 3-D.

3.3.2   Error of t-Cost Distances for Real Costs

We extend the definition of t-cost distance function by considering t as a real positive number. The functional form still remains the same. Here, we present a result of error estimate for real costs from [21].

Theorem 3.8 $\forall n \in N, \forall t \in \mathscr{R}^+$,
$$MRE(D_t^{(n)}) \leq (t - \lceil t \rceil).MRE(D_{\lfloor t \rfloor}^{(n)}) + (1 + \lceil t \rceil - t).MRE(D_{\lceil t \rceil}^{(n)}).$$

3.4   Error Analysis of Weighted t-Cost Distances (WtD)

By considering the MRE of t-cost norms, we express the MRE of a weighted-t-cost distance as follows [42].

Theorem 3.9 *From Theorem 3.6, the MRE of $WtD^{(n)}(\cdot; W)$ is given by the following expression:*

$$MRE(WtD^{(n)}(\cdot; W)) \leq \max_{1 \leq t \leq n} \left\{ \max \left(w_t \sqrt{t} - 1, 1 - \frac{w_t}{\max\left(1, \frac{\sqrt{n}}{t}\right)} \right) \right\}. \quad (3.9)$$

In the above equation, for a set of weights, the equality condition on the bound may not always hold. In weighted t-cost norm the maximally deviated t-cost norm (say at $t = i$) at a point in the space may be masked by the presence of closer contribution (to the Euclidean norm) from its other members in its family. Hence, for a set of weights W, its *strict upper bound* may be different from the RHS expression of Eq. (3.9). Let us denote the expression as the *simple upper bound (R_u)* of $MRE(WtD^{(n)}(\cdot; W))$.

Theorem 3.10 *The simple upper bound of $MRE(WtD^{(n)}(\cdot; W))$ (R_u) gets minimized by the following weight assignment:*

$$w_t = \begin{cases} \frac{2}{\sqrt{t} + \frac{t}{\sqrt{n}}} & t \leq \sqrt{n}. \\ \frac{2}{1 + \sqrt{t}} & \sqrt{n} \leq t \leq n. \end{cases} \quad (3.10)$$

With above weight assignment, the value of the simple upper bound of R_W^n becomes $\frac{\sqrt{n}-1}{\sqrt{n}+1}$.

Proof Let us minimize the MRE (say $MRE(w_t D_t^{(n)})$), for each $w_t D_t^{(n)}$ norm with respect to w_t. In this case, the MRE is given by the following: equation.

$$MRE(w_t D_t^{(n)}) = \max \left(w_t \sqrt{t} - 1, 1 - \frac{w_t}{\max\left(1, \frac{\sqrt{n}}{t}\right)} \right) \tag{3.11}$$

For $t \leq \sqrt{n}$, $MRE(w_t D_t^{(n)}) = \max \left(w_t \sqrt{t} - 1, 1 - \frac{w_t t}{\sqrt{n}} \right)$.

Above expression gets minimized when $w_t = \frac{2}{\sqrt{t} + \frac{t}{\sqrt{n}}}$ by solving the following equation:

$$w_t \sqrt{t} - 1 = 1 - \frac{w_t t}{\sqrt{n}}$$

For $t \geq \sqrt{n}$, $MRE(w_t D_t^{(n)}) = \max \left(w_t \sqrt{t} - 1, 1 - w_t \right)$.

Above expression gets minimized when $w_t = \frac{2}{1+\sqrt{t}}$ by solving the following equation:

$$w_t \sqrt{t} - 1 = 1 - w_t$$

The minimum value of $MRE(w_t D_t^{(n)})$ is given as

$$MRE(w_t D_t^{(n)}) = \begin{cases} \frac{2}{1+\sqrt{\frac{t}{n}}} - 1 & t \leq \sqrt{n}. \\ \frac{2}{1+\frac{1}{\sqrt{t}}} - 1 & \sqrt{n} \leq t \leq n. \end{cases} \tag{3.12}$$

Now, $MRE(WtD^{(n)}(\cdot; W)) = \max_{1 \leq t \leq n}(MRE(w_t D_t^{(n)}))$. The maximum value occurs at $t = 1$ or $t = n$ and it is found to be $\frac{\sqrt{n}-1}{\sqrt{n}+1}$. $\qquad\square$

3.5 Error Analysis of Chamfering Weighted Distances (CWD)

Several works have been reported [6, 7, 9–11, 80], on absolute error analysis of chamfer or weighted distances by considering the points on a finite grid. Though absolute error with respect to the corresponding Euclidean distance is unbounded, given a finite range of coordinate values the error is bounded and the maximum absolute error of a weighted distance function can be computed as a function of the size of the grid, as defined in the definition of MNAE (Definition 3.6). The optimization process is carried out in the continuous space for obtaining optimum values of weights of CWDs. Subsequently, their approximations with integral weights are considered. Let us illustrate the approach by considering the distance functions in 2-D. Subsequently, we review the results in higher dimensions.

3.5.1   2-D

Let us assume that the coordinate values range from 0 to M, and we would consider the problem of computing maximum absolute error (MAE) of a CWD given by $< \delta_1, \delta_2 >$. As the mask is central symmetric, it is sufficient to compute the maximum over a region $\mathscr{A} = \{(x, y)|0 \le x \le M, 0 \le y \le x\}$. In that case, the computation of MAE is performed as in the following:

$$MAE = \max_{(x,y) \in \mathscr{A}} \left\{ \left| \sqrt{x^2 + y^2} - \delta_1 x - (\delta_2 - \delta_1)y \right| \right\}$$

The above computation can be performed by setting x to M and finding out the value of y at which the following function $f(y)$ gets maximized.

$$f(y) = \left| \sqrt{M^2 + y^2} - \delta_1 M - (\delta_2 - \delta_1)y \right|$$

The derivative of the above function $f'(y)$ is 0 at $Y_0 = \frac{\delta_2 - \delta_1}{\sqrt{1 - (\delta_2 - \delta_1)^2}} M$. Hence the maximum value can occur either at extreme values of y (0 or M), or at $y = Y_0$. These correspond to three error terms such as $f(0) = (\delta_1 - 1)M$, $f(Y_0) = (\delta_1 - \sqrt{1 - (\delta_2 - \delta_1)^2})M$, and $f(M) = (\delta_2 - \sqrt{2})M$. Hence, $\max\{f(0), f(Y_0), f(M)\}$ would provide the MAE. Next minimization of MAE is carried out over the space of (δ_1, δ_2), such that $0 < \delta_1 \le \delta_2 \le 2\delta_1$ (conditions for the chamfer mask to induce norm). This computation gives us the optimum CWD for minimizing MNAE as stated in the following theorem:

Theorem 3.11 *In $\mathscr{R}^2$ a CWD represented by $< \delta_1, \delta_2 >$ minimizes the MNAE at $\delta_1 = \frac{1}{2}(\sqrt{2\sqrt{2} - 2} + 1) \approx 0.95509$, and $\delta_2 = \frac{1}{2}(\sqrt{2\sqrt{2} - 2} - 1) \approx 1.36930$ with the value of MNAE as $\frac{1}{2}(\sqrt{2\sqrt{2} - 2} - 1) \approx 0.04491$.*

We denote the above CWD as $CWD_{opt}^{(2)}$. If δ_1 is restricted to the value 1, which is the usual distance of a 4-neighbor from the point, the optimum CWD (denoted as $CWD_{opt*}^{(2)}$) providing minimum MNAE is given by the following theorem:

Theorem 3.12 *In $\mathscr{R}^2$ a CWD represented by $< 1, \delta >$ minimizes the MNAE at $\delta = \frac{1}{\sqrt{2}} + \sqrt{\sqrt{2} - 1} \approx 1.351$ with the value of MNAE as $\frac{1}{\sqrt{2}} - \sqrt{\sqrt{2} - 1} \approx 0.064$.*

Verwer [80] proposed to carry out the above optimization process over the points lying on the circle of radius 1. In this approach, the minimum and maximum CWDs of the points lying on the unit circle has been computed. The approach uses the geometry of a circle to compute the errors. This we discuss in the next chapter (Sect. 4.4.4.1).

Following this approach, he reported the MRE with a set of weights, very close to those obtained in [6]. He reported the optimum CWD as $< 0.9604, 1.3583 >$ with the MRE as 0.039566.

3.5.1.1 The Unbiased Mean Squared Error

Verwer [80] refined the error analysis by considering minimization of mean squared error of the deviations from Euclidean distance values over the points on the unit circle subject to that the average of the error becomes 0. The error is called *unbiased mean squared error* (UMSE). Given the polar coordinate representation of a point (x, y) as $(1, \phi)$, where $\sqrt{x^2 + y^2} = 1$ and $\phi = tan^{-1}\frac{y}{x}$, the optimization problem minimizes the following error:

$$UMSE(\delta_1, \delta_2) = \frac{4}{\pi} \int_0^{\frac{\pi}{4}} (1 - \delta_1 cos\phi - (\delta_2 - \delta_1)sin\phi)^2 d\phi \qquad (3.13)$$

so that

$$\int_0^{\frac{\pi}{4}} (1 - \delta_1 cos\phi - (\delta_2 - \delta_1)sin\phi)d\phi = 0 \qquad (3.14)$$

The solution of the above problem is stated in the following theorem [80].

Theorem 3.13 *In $\mathscr{R}^2$, the CWD providing the minimum unbiased mean squared error is given by $< 0.9481, 1.3408 >$ with the error value 0.0233.*

In our discussion, the above distance function is called $CWD_{umse}^{(2)}$.

3.5.1.2 Integral Approximation of Weights

In digital distances, it is often preferred to have integer weights in the definition of CWD. As $\frac{4}{3} = 1.33$ which is close to 1.351, the CWD $< 3, 4 >$ provides a close approximation of the Euclidean distance. However, the distance values are needed to be scaled by $\frac{1}{3}$ in this case. It is shown that with this distance function the MNAE becomes $(\sqrt{2} - \frac{4}{3}) \approx 0.08$. For $< 1, \sqrt{2} >$ (also called $CWD_{eu}^{(2)}$) the value of MNAE is $(1 - \sqrt{2\sqrt{2} - 2}) \approx 0.09$. A close approximation of this distance function with integral weights is $< 2, 3 >$ (as $\frac{3}{2}$ is close to $\sqrt{2}$) and the error value is $(1 - \frac{\sqrt{3}}{2}) \approx 0.13$. For the latter distance function, distance values should be scaled by $\frac{1}{2}$.

3.5.2   3-D

Borgefors has also carried out similar analysis in 3-D [6, 10]. In this case, the optimum CWD in the form of $< 1, \delta_2, \delta_3 >$ minimizing the MNAE is given by the following theorem:

Theorem 3.14 *In* $\mathscr{R}^3$ *a CWD represented by* $< 1, \delta_2, \delta_3 >$ *minimizes the MNAE at* $\delta_2 = \frac{1}{3}(\sqrt{3}+1+\sqrt{2\sqrt{3}-2}) \approx 1.314$, *and* $\delta_3 = \frac{1}{3}(2\sqrt{3}-1+2\sqrt{2\sqrt{3}-2}) \approx$ 1.628, *with the value of MNAE as* $\frac{1}{3}(\sqrt{3}+1-2\sqrt{2\sqrt{3}-2}) \approx 0.10$.

We refer to the above optimum distance function as $CWD^{(3)}_{opt*}$. A close approximation of this distance function with integral weights is $< 3, 4, 5 >$ as $\frac{4}{3} = 1.33$ and $\frac{5}{3} = 1.66$ are close values to δ_2 and δ_3 of $CWD^{(3)}_{opt*}$, respectively. The distance values are to be scaled by $\frac{1}{3}$. In this case, the MNAE is given by $\frac{1}{3}(1 - \frac{7}{\sqrt{3}}) \approx 0.12$.

When the weights are chosen as the Euclidean distances of respective neighbors, the CWD is referred to as $CWD^{(3)}_{eu}$ and is given as $< 1, \sqrt{2}, \sqrt{3} >$. For $CWD^{(3)}_{eu}$, the MNAE is given by approximately 0.15.

Verwer [80] also extended the approach of minimization of unbiased mean squared error (refer to Sect. 3.5.1.1) to 3-D. His results of the analysis are summarized in the following theorem:

Theorem 3.15 *In* $\mathscr{R}^3$, *the CWD providing the minimum unbiased mean squared error is given by* $< 0.8940, 1.3409, 1.5879 >$ *with the error value* 0.0233.

We refer to the above distance as $CWD^{(3)}_{umse}$.

3.5.3   4-D

In 4-D [11], the $CWD^{(4)}_{opt}$ is given by the following theorem.

Theorem 3.16 *In* $\mathscr{R}^4$ *a CWD represented by* $< 1, \delta_2, \delta_3, \delta_4 >$ *minimizes the MNAE at* $\delta_2 = \frac{1}{6}(\sqrt{32\sqrt{3}-50}+2\sqrt{3}+1) \approx 1.299$, *and* $\delta_3 = \frac{1}{3}(\sqrt{32\sqrt{3}-50}+2\sqrt{3}-1) \approx 1.598$, *and* $\delta_4 = \frac{1}{3}(\sqrt{32\sqrt{3}-50}-\sqrt{3}+5) \approx 1.866$, *with the value of MNAE as* $\frac{1}{3}(1+\sqrt{3}-\sqrt{32\sqrt{3}-50}) \approx 0.13$.

A close approximation with integral weights of the above function is given by $< 3, 4, 5, 6 >$ with the value of MNAE as $1 - \frac{\sqrt{3}}{6} \approx 0.18$ if the distance is scaled by $\frac{1}{3}$. A slightly better approximation is given by the CWD $< 7, 9, 11, 13 >$ leading with MNAE as $\sqrt{3} - \frac{11}{7} \approx 0.16$, when the distances are divided by 7. It may be noted that for $CWD^{(4)}_{eu}$ the MNAE is given by $(1 - \sqrt{2\sqrt{6}+4\sqrt{3}+2\sqrt{2}-14}) \approx 0.19$.

3.5.4 N-D

In an arbitrary n-dimensional real space, Barni et al. [2] derived a CWD which provides the minimum MRE. The results derived by them are summarized in the following theorem. The proofs of several lemmas leading to this theorem are provided in [2].

Theorem 3.17 *In $\mathscr{R}^n$ the weights of CWD $< \delta_1, \delta_2, \ldots, \delta_n >$ providing minimum MRE are given by the following expression:*

$$\delta_i = \kappa \sqrt{i} \, for \, 1 \leq i \leq n \tag{3.15}$$

where κ is given by

$$\kappa = \frac{2}{1 + \sqrt{\sum_{i=1}^{n}(\sqrt{i} - \sqrt{i-1})^2}} \tag{3.16}$$

The MRE of this distance function is given by $1 - \kappa$.

We refer to the above CWD as $CWD_{euopt}^{(n)}$.

3.5.5 Error Analysis for CMIDs with Larger Neighborhoods

For CMIDs, whose closed form expressions for distance values are obtained (refer to Sect. 2.3.1), techniques similar to finding optimum values of weights for CWDs are adopted. The results from [8, 76, 80], are summarized in Table 3.2.

Table 3.2 Optimum CMIDs in 2D and 3D

Chamfer mask	Optimum parameters	Minimum error
2D		
$M_{55}^{(2)}(a, b, c)$	$a_{opt} = 0.98587, 1.40018 \leq b_{opt} \leq 1.42178, c_{opt} = 2.20780$	MNAE: 0.01413
$M_{55}^{(2)}(a, b, c)$	$a_{opt} = 0.9801, b_{opt} = 1.4060, c_{opt} = 2.2044$	UMSE: 0.00706
3D		
$M_{555}^{(3)}(a, b, c, e)$	$a_{opt} = 0.95446, b_{opt} = 1.36868, c_{opt} = 1.68651,$ $e_{opt} = 2.37085$	MNAE: 0.04554
$M_{555}^{(2)}(a, b, c, d, e)$	$b_{opt} = 1.37755, c_{opt} = 1.69538, e_{opt} = 2.40252$ $0.96333 \leq a_{opt} \leq 0.99179, \text{ at } a_{opt} = 0.96333,$ $2.16459 \leq d_{opt} \leq 2.28307 \text{ at } a_{opt} = 0.99179,$ $2.19302 \leq d_{opt} \leq 2.19534$	MNAE: 0.036667
$M_{555}^{(2)}(a, b, c, d, e, f)$	$a_{opt} = 0.97131, b_{opt} = 1.38553, c_{opt} = 1.70336,$ $d_{opt} = 2.17869, e_{opt} = 2.41476, 3.00712 \leq f_{opt} \leq 3.04444$	MNAE: 0.02869
$M_{555}^{(2)}(a, b, c, d, e, f)$	$a_{opt} = 0.9556, b_{opt} = 1.3956, c_{opt} = 1.7257, d_{opt} = 2.1830,$ $e_{opt} = 2.3885, f_{opt} = 2.9540$	UMSE: 0.00782

3.5.5.1 Integral Approximation of Optimum Parameters

For computing with integral weights in $\mathscr{L}^2$ and $\mathscr{L}^3$, the optimum distance functions
are approximated by integral weights. The distance values scaled by a factor due to
integral approximation are found to be close to the corresponding values of Euclidean
norm. A list of such distance functions with the error values is provided in Table 3.3
from [8, 76, 80].

Table 3.3 Integral CMIDs for chamfer masks of sizes 5×5 in 2D and $5 \times 5 \times 5$ in 3D

Integral parameters	Scale	Error
2D		
$a = 4, b = 6, c = 9$	0.2414	MNAE: 0.03448
$a = 4, b = 6, c = 9$	0.2427	UMSE: 0.0155
$a = 5, b = 7, c = 11$	0.1998	MNAE: 0.01915
$a = 5, b = 7, c = 11$	0.1992	UMSE: 0.0105
$a = 6, b = 8, c = 13$	0.1714	MNAE: 0.04312
$a = 7, b = 10, c = 16$	0.2450	MNAE: 0.02000
$a = 9, b = 13, c = 20$	0.1100	**MNAE: 0.01835**
$a = 9, b = 13, c = 20$	0.1094	UMSE: 0.0085
$a = 9, b = 13, c = 21$	0.1083	MNAE: 0.02875
$a = 16, b = 23, c = 36$	0.0612	**UMSE: 0.0071**
3D		
$a = 3, b = 4, c = 5, e = 7$	$\frac{1}{3}$	MNAE: 0.0809
$a = 3, b = 4, c = 5, d = 6, e = 7, f = 9$	0.3420	UMSE: 0.0250
$a = 4, b = 6, c = 7, d = 9, e = 10, f = 12$	0.2413	UMSE: 0.0123
$a = 5, b = 7, c = 9, d = 11, e = 12, f = 15$	0.1970	UMSE: 0.0116
$a = 7, b = 10, c = 12, d = 16, e = 17$	$\frac{1}{7}$	MNAE: 0.0754
$a = 7, b = 10, c = 12, d = 16, e = 17, f = 21$	$\frac{1}{7}$	MNAE: 0.0524
$a = 8, b = 11, c = 14, d = 18, e = 20$	$\frac{1}{8}$	MNAE: 0.0653
$a = 9, b = 13, c = 16, d = 20, e = 22, f = 27$	0.1085	UMSE: 0.0098
$a = 10, b = 14, c = 17, d = 22, e = 24$	$\frac{1}{10}$	MNAE: 0.0519
$a = 10, b = 14, c = 17, d = 22, e = 24, f = 30$	$\frac{1}{10}$	MNAE: 0.0408
$a = 11, b = 15, c = 19, e = 27$	$\frac{1}{11}$	MNAE: 0.0754
$a = 13, b = 18, c = 22, d = 29, e = 31$	$\frac{1}{13}$	MNAE: 0.0397
$a = 13, b = 18, c = 22, d = 29, e = 31, f = 38$	$\frac{1}{13}$	MNAE: 0.0397
$a = 14, b = 19, c = 24, e = 34$	$\frac{1}{14}$	MNAE: 0.0687
$a = 17, b = 23, c = 29, e = 41$	$\frac{1}{17}$	MNAE: 0.0662
$a = 17, b = 25, c = 31, d = 39, e = 43, f = 53$	0.0557	**UMSE: 0.0082**
$a = 20, b = 27, c = 34, e = 48$	$\frac{1}{20}$	MNAE: 0.0646
$a = 23, b = 32, c = 39, d = 51, e = 55, f = 68$	$\frac{1}{23}$	**MNAE: 0.0395**

3.6 Error Analysis of Octagonal Distances

In [20], error analysis of simple octagonal distances with the neighborhood sequence $B = \{b(1), b(2), \ldots, b(p)\}$ in $\mathscr{Z}^2$ has been carried out. With an involved analysis, Das [20] could derive the expressions for the MNAE and the MRE of a simple octagonal distance $d_B^{(2)}(\cdot|q, p)$, with $q = f(p) = \sum_{i=1}^{p} b(i)$ (refer to Theorem 2.13). Though the analysis was carried out for a finite grid, the results are also applicable for the whole $\mathscr{Z}^2$. In following two theorems, we state these results.

Theorem 3.18

$$MNAE(d_B^{(2)}(\cdot|q, p)) = \max \left\{ \sqrt{(1 + (m-1)^2)} - 1, \left| \frac{2}{m} - \sqrt{2} \right| \right\}$$

where $m = \frac{q}{p}$, and $1 \leq m \leq 2$.

Theorem 3.19

$$MRE(d_B^{(2)}(\cdot|q, p)) = \max \left\{ 1 - \frac{1}{\sqrt{(1 + (m-1)^2)}}, \left| 1 - \frac{\sqrt{2}}{m} \right| \right\}$$

where $m = \frac{q}{p}$, and $1 \leq m \leq 2$.

The optimum value of m, for which the MNAE is minimized can be obtained by solving the following equation:

$$\sqrt{(1 + (m-1)^2)} - 1 = \left| \frac{2}{m} - \sqrt{2} \right| \qquad (3.17)$$

The solutions of the above equation are numerically found to be at two values of m: $m = 1.355$ and $m = 1.6076$. By comparing the values of MNAE at these two values, optimum m is chosen as 1.355 with $MNAE = 0.0613$. A close approximation to this optimum m could be obtained by the neighborhood sequence $B = \{1, 1, 2\}$, where $m = \frac{4}{3} \approx 1.33$. In the same way, the optimum m minimizing the MRE is found at $m = 1.3420$ with the MRE value 0.0538. The optimum value of m for the MRE is quite close to that of MNAE and the same neighborhood sequence $B = \{1, 1, 2\}$ is a good sequence to closely approximate this optimum value of m. We should note that there are many other neighborhood sequences which provide close approximations of these values of m.

Das [20] also derived an expression for *average normalized absolute error* (ANAE), which is computed on a bounded region of grid size M. The normalized average error is found to be a continuous function of m. Its average is computed by integrating the function over the set of points within that region and then dividing the result of integration by the area of the region. Let us denote the error as $ANAE(d_B^{(2)}(\cdot|q,p))$ for a simple octagonal distance $d_B^{(2)}(\cdot|q,p)$. Similarly average of relative errors (ARE) is also computed for $d_B^{(2)}(\cdot|q,p)$, and it is denoted as $ARE(d_B^{(2)}(\cdot|q,p))$. In the following theorems, we state the results of the analysis done by Das [20].

Theorem 3.20 *The ANAE of $d_B^{(2)}(\cdot|q,p)$ is given by the following expression:*
$ANAE(d_B^{(2)}(\cdot|q,p))$

$$
\begin{aligned}
&= \tfrac{1}{3}\big((2-\sqrt{2}) - ln(\sqrt{2}+1) - \tfrac{m^2-2+4t(m)+4t^2(m)}{m} + \\
&\quad 2t(m)\sqrt{1+t^2(m)} + 2ln(t(m)) + \sqrt{1+t^2(m)}\big) \quad 1 \le m \le \sqrt{2} \qquad (3.18)\\
&= \tfrac{1}{3}\big((2+\sqrt{2}) + ln(\sqrt{2}+1) - \tfrac{m^2+4}{m}\big) \qquad\qquad\qquad \sqrt{2} \le m \le 2
\end{aligned}
$$

where $m = \frac{q}{p}$, $1 \le m \le 2$, and $t(m) = \frac{1-m\sqrt{2-m^2}}{m^2-1}$.

Theorem 3.21 *The ARE of $d_B^{(2)}(\cdot|q,p)$ is given by the following expression:*
$ARE(d_B^{(2)}(\cdot|q,p))$

$$
\begin{aligned}
&= (2t(m)-1) - (1-\tfrac{1}{m})ln(m-1+\sqrt{1+(m-1)^2}) \\
&\quad + \tfrac{\sqrt{1+(m-1)^2}}{m} + \tfrac{1}{m}(\sqrt{2}+ln(\sqrt{2}+1)) \\
&\quad - \tfrac{2}{m}(ln(t(m)) + \sqrt{1+t^2(m)}) + \sqrt{1+t^2(m)} \qquad 1 \le m \le \sqrt{2} \qquad (3.19)\\
&= 1 - (1-\tfrac{1}{m})(ln(m-1+\sqrt{1+(m-1)^2})) \\
&\quad - \tfrac{1}{m}(\sqrt{2}+ln(\sqrt{2}+1)) + \tfrac{\sqrt{1+(m-1)^2}}{m} \qquad\qquad \sqrt{2} \le m \le 2
\end{aligned}
$$

where $m = \frac{q}{p}$, $1 \le m \le 2$, and $t(m) = \frac{1-m\sqrt{2-m^2}}{m^2-1}$.

It is interesting to note that by numerical methods, Das [20], found that minimum values of both ANAE and ARE occur at $m = 1.400001$. At this optimum value of m, the values of ANAE and ARE are 0.015950, and 0.021651, respectively. The neighborhood sequence $B = \{1, 1, 2, 1, 2\}$ provides a close approximation of this value of m. For this sequence, the value of m is $\frac{7}{5} = 1.40$.

Some of the good simple octagonal distances are listed in Table 3.4, with their values of MNAE, MRE, ANAE, and ARE from [20].

Table 3.4 Typical good simple octagonal distances

Distance functions	m	MNAE	MRE	ANAE	ARE
$d_{12}^{(2)}$	$\frac{3}{2}$	0.118	0.106	0.043	0.055
$d_{112}^{(2)}$	$\frac{4}{3}$	0.086	**0.061**	0.026	0.032
$d_{1112}^{(2)}$	$\frac{5}{4}$	0.186	0.131	0.056	0.068
$d_{11212}^{(2)}$	$\frac{7}{5}$	**0.077**	0.072	**0.016**	**0.022**

In [83], it is shown that the relative error of a HOD is bounded, while the absolute error is not. Error analysis of HODs for higher dimension is quite challenging, and not many works have been reported following an analytical approach. However, using geometric approaches, a few such efforts exist, which are discussed in the next chapter.

3.7   Empirical Computation of Errors

Sometimes it is quite challenging to derive a closed form analytical expression for computing the errors as defined above. In such cases, indicative values, computed empirically, may be used over a finite grid. In some works, for evaluating the closeness of digital distances to Euclidean metrics, these empirical error measures are used [20, 42–44, 47], for studying the performances of distance functions in lower dimensional spaces such as in $\mathscr{Z}^2$, $\mathscr{Z}^3$, $\mathscr{Z}^4$, etc.

3.7.1   Empirical Measures

Different analytical errors, such as MNAE and MRE as discussed before, could be computed empirically. Moreover, the average of normalized absolute error (ANAE) and relative errors (ARE) can also be computed empirically. Definitions of a few of these empirical error measures are provided below.

Definition 3.8 A few *empirical analytical errors* for a distance function d in $\mathscr{Z}_M{}^n$ are defined as follows:

1. The *empirical normalized average error* (ENAE):

$$\mathscr{E}_{avg}^{(n)}(d) = \frac{1}{M} \left\{ \frac{\sum_{i_1=0}^{M} \sum_{i_2=0}^{i_1} \cdots \sum_{i_n=0}^{i_{n-1}} |E^{(n)}(i_1, i_2, \ldots i_n) - d(i_1, i_2, \ldots i_n)| \, \Phi(i_1, i_2, \ldots i_n, M)}{\sum_{i_1=0}^{M} \sum_{i_2=0}^{i_1} \cdots \sum_{i_n=0}^{i_{n-1}} \Phi(i_1, i_2, \ldots i_n, M)} \right\}$$

2. The *empirical maximum normalized absolute error* (EMNAE):

$$\mathscr{E}_{max\_diff}^{(n)}(d) = \frac{1}{M} \left\{ \max_{0 \leq i_n \leq i_{n-1} \ldots \leq i_1 \leq M} |E^{(n)}(i_1, i_2, \ldots i_n) - d(i_1, i_2, \ldots i_n)| \, \Phi(i_1, i_2, \ldots i_n, M) \right\}$$

3. The *empirical average relative error* (EARE):

$$
\mathscr{R}_{avg}^{(n)}(d) = \frac{\left\{ \sum\limits_{i_1=0}^{M} \sum\limits_{i_2=0}^{i_1} \cdots \sum\limits_{i_n=0}^{i_{n-1}} \frac{|E^{(n)}(i_1,i_2,\ldots i_n) - d(i_1,i_2,\ldots i_n)|}{E^{(n)}(i_1,i_2,\ldots i_n)} \Phi(i_1, i_2, \ldots i_n, M) \right\}}{\sum\limits_{i_1=0}^{M} \sum\limits_{i_2=0}^{i_1} \cdots \sum\limits_{i_n=0}^{i_{n-1}} \Phi(i_1, i_2, \ldots i_n, M)}
$$

4. The *empirical maximum relative error* (EMRE):

$$
\mathscr{R}_{max}^{(n)}(d) = \max_{0 \leq i_n \leq i_{n-1} \cdots \leq i_1 \leq M} \frac{|E^{(n)}(i_1, i_2, \ldots i_n) - d(i_1, i_2, \ldots i_n)|}{E^{(n)}(i_1, i_2, \ldots i_n)} \Phi(i_1, i_2, \ldots i_n, M)
$$

where

$$
\Phi(i_1, i_2, \ldots i_n, M) = \begin{cases} 1, & \text{if } E^{(n)}(i_1, i_2, \ldots i_n) \leq M, \\ 0, & \text{Otherwise.} \end{cases} \tag{3.20}
$$

We note that the computation of the above errors is restricted within the Euclidean hypersphere of radius M. This is to remove the biases imposed by some of the directions such as diagonals of hypercubes, which have more number of points as observed in [80]. The values of M are usually kept at 512 in 2-D and 3-D, and 256 in 4-D in reported experimentation [43].

In the literature, empirical errors of distance functions are mostly reported for distances in 2-D and 3-D. This is due to the fact that as the dimension increases, the empirical estimates of these errors increasingly deviate from true values in small sized finite grids. Hence, to make the observation robust the number of points should be greatly increased [13]. This may require a large computation and imposes a severe bottleneck for extending this method to distances in moderate to high dimensional spaces. Moreover, these estimates are biased because the errors are only computed on points lying along a few discrete directions. Usually, they are found to be optimistic as observed in [14]. To overcome it partly, there are empirical error measures [14], which compute those errors on a finite set of sampled points on unit hypersphere of the distance function. For a realistic estimate, points in the space should be sampled with finer resolution and the computation of error should be iterated with an increasing number of samples till it converges. In the next chapter, we discuss the use of such empirical errors.

3.7.2   Empirical Analytical Errors of a Few Distance Functions

For computing ENAE and EMNAE in 2-D and 3-D, usually the value of M is taken as 512. This is primarily motivated by the fact that in many applications sizes of digital images in 2-D or 3-D are well within these ranges (i.e 512 × 512 in 2-D and 512 × 512 × 512 in 3-D). Also, the error values are found to be stabilized over this set

Table 3.5 Relative performances of good 2D octagonal distances and good weighted distances considering ENAE and EMNAE measures

Octagonal distances	$\mathcal{E}_{avg}^{(2)}$	$\mathcal{E}_{max\_diff}^{(2)}$	Weighted distances	$\mathcal{E}_{avg}^{(2)}$	$\mathcal{E}_{max\_diff}^{(2)}$
$d_{12}^{(2)}$	0.026	0.105	$d_{<2,3>}^{(2)}$	0.058	0.118
$d_{112}^{(2)}$	**0.014**	0.062	$d_{<3,4>}^{(2)}$	0.023	**0.057**
$d_{1112}^{(2)}$	0.030	0.133	$d_{<5,7>}^{(2)}$	0.033	0.077
$d_{122}^{(2)}$	0.041	0.168	$d_{<8,11>}^{(2)}$	0.028	0.068

Table 3.6 Relative performances of good HODs and CWDs in 3D considering ENAE and EMNAE measures

Octagonal distances	$\mathcal{E}_{avg}^{(3)}$	$\mathcal{E}_{max\_diff}^{(3)}$	Weighted distances	$\mathcal{E}_{avg}^{(3)}$	$\mathcal{E}_{max\_diff}^{(3)}$
$d_{113}^{(3)}$	**0.022**	0.095	$d_{<3,4,5>}^{(3)}$	0.042	0.106
$d_{11123}^{(3)}$	0.026	**0.087**	$d_{<8,11,13>}^{(3)}$	0.043	0.097
$d_{1111223}^{(3)}$	0.031	0.106	$d_{<13,17,22>}^{(3)}$	0.043	0.113
$d_{12}^{(3)}$	0.044	0.156	$d_{<13,17,23>}^{(3)}$	0.054	0.138
$d_{122}^{(3)}$	0.043	0.168	$d_{<16,21,27>}^{(3)}$	0.042	0.112
$d_{13}^{(3)}$	0.068	0.183	$d_{<16,21,28>}^{(3)}$	0.052	0.132

Table 3.7 EARE and EMRE of distances in different dimensions

Distance (2D)	$\mathcal{R}_{avg}^{(n)}$	$\mathcal{R}_{max}^{(n)}$	Distance (3D)	$\mathcal{R}_{avg}^{(n)}$	$\mathcal{R}_{max}^{(n)}$	Distance (4D)	$\mathcal{R}_{avg}^{(n)}$	$\mathcal{R}_{max}^{(n)}$
$CWD_{eu}^{(2)}$	0.024	**0.040**	$CWD_{eu}^{(3)}$	0.030	**0.060**	$CWD_{eu}^{(4)}$	**0.034**	**0.074**
$<3,4>$	0.034	0.057	$<3,4,5>$	0.056	0.106	$<3,4,5,6>$	0.085	0.155
$CWD_{opt}^{(2)}$	0.034	0.057	$CWD_{opt}^{(3)}$	0.032	0.074	$CWD_{opt}^{(4)}$	0.085	0.155
$CWD_{opt*}^{(2)}$	0.037	0.060	$CWD_{opt*}^{(3)}$	0.048	0.094	$CWD_{opt*}^{(4)}$	0.052	0.118
$CWD_{umse}^{(2)}$	**0.020**	0.052	$CWD_{umse}^{(3)}$	**0.019**	0.106			
$<1,1>$	0.100	0.293	$<1,1,1>$	0.169	0.423	$<1,1,1,1>$	0.221	0.500
$<1,2>$	0.273	0.414	$<1,2,3>$	0.499	0.732	$<1,2,3,4>$	0.695	1.000
{112}	0.028	0.060	{113}	0.029	0.095	{114}	0.025	0.134
{1112}	0.058	0.131	{1111223}	0.042	0.104	{1124}	0.040	0.147
$WtD_{isr}^{(2)}$	**0.024**	0.076	$WtD_{isr}^{(3)}$	0.041	0.113	$WtD_{isr}^{(4)}$	0.037	0.138

of points. In Tables 3.5 and 3.6, some of the good digital distances from the families of HODs and CWDs are illustrated with their ENAE and EMNAE values from the Euclidean metrics. In Table 3.7, EARE and EMRE values of different distance functions in 2-D, 3-D, and 4-D are also shown [43]. In 4-D the value of M is taken as 256.

3.8   Concluding Remarks

From analytical approaches, we get a few good digital distances which are close to Euclidean metrics in respective spaces. For example, in 2-D, the octagonal distances $d_{112}^{(2)}$ and $d_{1112}^{(2)}$ have low errors, and out of CWDs $d_{<3,4>}^{(2)}$, and $d_{<5,7>}^{(2)}$ have good performances. If we use larger chamfering masks, we get further reduction in errors as it is observed that $d_{\mathscr{M}_{55}(5,7,11)}$, and $d_{\mathscr{M}_{55}(9,13,20)}$, have a very low MNAE, as well as a low UMSE. The UMSE of the CMID $d_{\mathscr{M}_{55}(16,23,36)}$ is very close to the optimal value. In 3-D and 4-D also, we get a few such distances. In 3-D the HODs $d_{113}^{(3)}$ and $d_{1111223}^{(3)}$ are good examples. Among CWDs $d_{<3,4,5>}^{(3)}$ may be recommended. In 4-D, we have distances such as $d_{114}^{(4)}, d_{1124}^{(4)}$, and $d_{<3,4,5,6>}^{(4)}$ as good replacement for the corresponding Euclidean metric. With larger neighborhoods, the CMIDs such as $d_{\mathscr{M}_{555}(17,25,31,39,43,53)}$, and $d_{\mathscr{M}_{555}(23,32,39,51,55,68)}$ are found to provide good approximation of Euclidean distances in 3-D. As we consider distances in a higher dimension, it is difficult to derive analytical error estimates. In such cases, we may compute empirical errors. But due to the explosion of size of space in a high dimension, it is not feasible to adopt this approach in a very high dimensional space. In the next chapter, we discuss geometric approaches of studying proximity to Euclidean metrics, which provide another perspective in studying errors in low and high dimensional spaces.

Chapter 4
Error Analysis: Geometric Approaches

In this chapter, we discuss geometric approaches for analyzing errors of distances from Euclidean metrics. In this approach, hyperspheres of a distance function are studied and their properties such as hypersurface area, hypervolume, shape, etc., are compared with those of Euclidean hyperspheres in respective dimensions. In our discussion, sometimes we refer to hyperspheres as disks of a distance function.

4.1 Hyperspheres of Digital Distances

A hypersphere $\mathcal{H}$ of a given radius r around the origin $\bar{0}$ is the set of points whose distances from $\bar{0}$ are less than or equal to r. The set of points which are exactly at distance r from $\bar{0}$ forms the surface $\mathcal{S}$ of the hypersphere. In our discussion, we assume that the center of a hypersphere is origin 0, if it is not explicitly mentioned.

The volume of the hypersphere is given by the number of points in $\mathcal{H}$, and that of $\mathcal{S}$ provides the surface area.

Definition 4.1 $\mathcal{S}(d(\cdot); r)$ is the *Hypersurface* of radius r in n-D for the distance function $d(\cdot)$. It is the set of n-D grid points that lie *exactly* at a distance r, $r \geq 0$, from the origin when $d(\cdot)$ is used as the distance.

$$\mathcal{S}(d(\cdot); r) = \{\bar{x} : \bar{x} \in \mathcal{Z}^n, d(\bar{x}) = r\}$$

The *Surface Area surf*$(d(\cdot); r) = |\mathcal{S}(d(\cdot); r)|$ of a hypersurface $\mathcal{S}(d(\cdot); r)$ is defined as the number of points in $\mathcal{S}(d(\cdot); r)$.

In the digital space, *surf*$(d(\cdot); r)$ often is a polynomial of r of degree $n - 1$ with rational coefficients.

Definition 4.2 $\mathcal{H}(d(\cdot); r)$ is the *Hypersphere* of radius r in n-D for the distance function $d(\cdot)$. It is the set of n-D grid points that lie *within* a distance r, $r \geq 0$, from the origin.

$$\mathcal{H}(d(\cdot); r) = \{\bar{x} : \bar{x} \in \mathcal{Z}^n, 0 \leq d(\bar{x}) \leq r\}$$

The *Volume* $vol(d(\cdot); r) = |\mathcal{H}(d(\cdot); r)|$ of a hypersphere $\mathcal{H}(d(\cdot); r)$ is defined as the number of points in $\mathcal{H}(d(\cdot); r)$.

Definition 4.3 A hypersphere of unit radius, $\mathcal{H}(d(\cdot); 1)$, is called *unit hypersphere*.

In the digital space, $vol(d(\cdot); r)$ often is a polynomial of r of degree n with rational coefficients. A hypersphere of radius r is composed of its hypersurfaces of radii smaller or equal to r, as it is noted below.

$$\mathcal{H}(d(\cdot); r) = \bigcup_{s=0}^{r} \mathcal{S}(d(\cdot); s) \quad and \quad vol(d(\cdot); r) = \sum_{s=0}^{r} surf(d(\cdot); s)$$

We also define a feature related to the shape of a hypersphere.

Definition 4.4 The *Shape Feature* $\psi_n(d(\cdot))$ of a Hypersphere $\mathcal{H}$ in n-D is defined as

$$\psi_n(d(\cdot)) = \lim_{r \to \infty} \frac{(surf(d(\cdot); r)^n}{(vol(d(\cdot); r))^{n-1}}$$

The definition of shape feature $\psi_n(d(\cdot))$ makes it a dimension-less quantity. Though it is defined for any arbitrary dimension, it is more used to study shapes of hyperspheres in 2-D and 3-D, as the shapes are physically visualized in these dimensions. We should also note that, if $vol(d(\cdot); r)$ and $surf(d(\cdot); r)$ contain only the term for the highest power of r, the limit in the definition of shape becomes redundant. In Table 4.1, we enumerate the vertices, surface area, and volume of disks of m-neighbor distances of radius r in 2-D and 3-D from [29].

In the subsequent sections, we discuss different properties of hyperspheres of different types of digital distances. In some cases, there are closed-form expressions

Table 4.1 Properties of hyperspheres of m-neighbor distances in 2-D and 3-D

Distance	Vertices	Perimeter/Surface area	Area/Volume
$d_1^{(2)}$	$\{(\pm r, 0), (0, \pm r)\}$	$4r$	$2r^2 + 2r + 1$
$d_2^{(2)}$	$\{(\pm r, \pm r)\}$	$8r$	$4r^2 + 4r + 1$
$d_1^{(3)}$	$\{(\pm r, 0, 0), (0, \pm r, 0), (0, 0, \pm r)\}$	$24r^2 + 2$	$18r^3 + 12r^2 + 6r + 1$
$d_2^{(3)}$	$\{(\pm r, \pm r, 0), (\pm r, 0, \pm r), (0, \pm r, \pm r)\}$	$20r^2 - 4r + 2$	$\frac{20}{3}r^3 + 8r^2 + \frac{10}{3}r + 1$
$d_3^{(3)}$	$\{(\pm r, \pm r, \pm r)\}$	$4r^2 + 2$	$\frac{4}{3}r^3 + 2r^2 + \frac{8}{3}r + 1$

for their volumes and surfaces. Using volume and surface error measures with respect to an Euclidean hypersphere of the same radius, we also study how well these digital hyperspheres approximate the Euclidean hypersphere.

Before defining a vertex of a hypersphere, let us define 2^n *Symmetry Function* $\phi(\bar{x})$, where $\bar{x}$ is a point in $\mathscr{L}^n$ or $\mathscr{R}^n$.

Definition 4.5 $\phi(\cdot)$ is defined as the 2^n *Symmetry Function* of an n-D point. That is, given $\bar{x} \in \mathscr{L}^n$, and $x_i \geq 0, \forall i$, $\phi(\bar{x})$ gives the set of points in $\mathscr{L}^n$ obtained by reflections and permutations of $\bar{x}$.

For example, if $\bar{x} = (1, 3, 1)$, we get the set of 2^3-symmetric points as

$$\phi(\bar{x}) = \{(\pm 3, \pm 1, \pm 1), (\pm 1, \pm 3, \pm 1), (\pm 1, \pm 1, \pm 3)\}$$

In the same way, the definition of $\phi(\cdot)$ is also extended in $\mathscr{R}^n$.

Definition 4.6 A polytope in $\mathscr{R}^n$ is called 2^n *Symmetry Polytope* if its vertices preserve the property of 2^n symmetry, i.e., for every vertex $\bar{x}$, all the points in $\phi(\bar{x})$ are also its vertices.

For examples, all regular polytopes being centered at the origin have 2^n symmetry. In the present context, hyperspheres of all the distances that we would be considering are centered at the origin of the space, and they are 2^n *Symmetry Convex Polytopes*. These definitions are also trivially extended in a real space.

4.1.1 Euclidean Hyperspheres

To compare the digital hyperspheres of various neighborhood sets with the hyperspheres $\mathscr{H}(E^{(n)}; r)$, $r \geq 0$, of the Euclidean norm in n-D, we note the following:

Definition 4.7 The *Euclidean surface area*, *volume*, and *shape feature* measures are given as

$$
\begin{aligned}
\text{Volume:} \quad & V_n^E(r) & = & \ \{\bar{x} : \bar{x} \in \mathscr{L}^n, E^{(n)}(\bar{x}) \leq r\} & = & \ L_n r^n \\
\text{Surface Area:} \quad & S_n^E(r) & = & \ \{\bar{x} : \bar{x} \in \mathscr{L}^n, E(n)(\bar{x}) = r\} & = & \ \frac{d}{dr} V_n^E(r) = n L_n r^{n-1} \\
\text{Shape Feature:} \quad & \psi_n^E & = & \ \frac{(S_n^E(r))^n}{(V_n^E(r))^{n-1}} & = & \ n^n L_n
\end{aligned}
$$

where

$$L_n = \frac{\pi^{\lfloor n/2 \rfloor}}{\prod\limits_{i=0}^{\lceil n/2 \rceil - 1} \left(\frac{n}{2} - i\right)}$$

Above expressions can also be written by separately showing their algebraic forms for even and odd-dimensional spaces. These are also shown below.

$$V_n^E(r) = \begin{cases} \frac{\pi^k}{k!} r^{2k} & \text{for } n = 2k, \\ \frac{k!}{(2k+1)!} 2^{2k+1} \pi^k r^{2k+1} & \text{for } n = 2k + 1. \end{cases} \tag{4.1}$$

$$S_n^E(r) = \begin{cases} 2\frac{\pi^k}{(k-1)!} r^{2k-1} & \text{for } n = 2k, \\ \frac{k!}{(2k)!} 2^{2k+1} \pi^k r^{2k} & \text{for } n = 2k + 1. \end{cases} \tag{4.2}$$

The comparison is quantified in terms of the following error measures:

Definition 4.8 The *Error Measures* between Euclidean and digital hyperspheres are defined as

$$\begin{array}{llll} \textit{Surface Error:} & E_S^{(n)}(d(\cdot)) & = & \lim_{r\to\infty} \left| \frac{S_n^E(r) - surf(d(\cdot); r)}{r^{n-1}} \right| \\[2mm] \textit{Volume Error:} & E_V^{(n)}(d(\cdot)) & = & \lim_{r\to\infty} \left| \frac{V_n^E(r) - vol(d(\cdot); r)}{r^n} \right| \\[2mm] \textit{Shape Feature Error:} & E_\psi^{(n)}(d(\cdot)) & = & \left| \psi_n^E - \psi_n(d(\cdot)) \right| \end{array}$$

There are also other approaches for expressing geometric errors. One of them is based on the following identities related to volume, surface, and shape features of a Euclidean hypersphere:

$$\pi = \begin{cases} \left(\frac{k! V_n^E(r)}{r^n} \right)^{\frac{1}{k}} = \left(\frac{(k-1)! S_n^E(r)}{2r^{(n-1)}} \right)^{\frac{1}{k}} = \left(\frac{k! \psi_n^E}{n^n} \right)^{\frac{1}{k}} & \text{for } n = 2k, \\[3mm] \left(\frac{n! V_n^E(r)}{k! 2^n r^n} \right)^{\frac{1}{k}} = \left(\frac{(n-1)! S_n^E(r)}{k! 2^n r^{(n-1)}} \right)^{\frac{1}{k}} = \left(\frac{n! \psi_n^E}{k! 2^n n^n} \right)^{\frac{1}{k}} & \text{for } n = 2k + 1 \end{cases} \tag{4.3}$$

In [43], above identities are used to define a set of geometric errors by computing deviation from the value of π by using the same algebraic forms of expressions in Eq. (4.3), where Euclidean quantities are replaced by the respective measures of a hypersphere. Let us call these errors π-*errors*, namely *Volume-π*, *Surface-π*, and *Shape-π* errors. They are defined as follows:

Definition 4.9 The π-*error measures* between Euclidean and digital hyperspheres are defined as

Surface-π Error:

$$E_{s\pi}^{(n)}(d(\cdot)) = \begin{cases} \left| \pi - \left(\frac{(k-1)! surf(d(\cdot); r)}{2r^{(n-1)}} \right)^{\frac{1}{k}} \right| & \text{for } n = 2k, \\[3mm] \left| \pi - \left(\frac{(n-1)! surf(d(\cdot); r)}{k! 2^n r^{(n-1)}} \right)^{\frac{1}{k}} \right| & \text{for } n = 2k + 1 \end{cases} \tag{4.4}$$

Volume-π Error

$$E_{v\pi}^{(n)}(d(\cdot)) = \begin{cases} \left| \pi - \left(\frac{k! vol(d(\cdot); r)}{r^n} \right)^{\frac{1}{k}} \right| & \text{for } n = 2k, \\[3mm] \left| \pi - \left(\frac{n! vol(d(\cdot); r)}{k! 2^n r^n} \right)^{\frac{1}{k}} \right| & \text{for } n = 2k + 1 \end{cases} \tag{4.5}$$

Shape-π Error

$$E_{\psi\pi}^{(n)}(d(\cdot)) = \begin{cases} \left| \pi - \left(\frac{k!\psi_n(d(\cdot))}{n^n} \right)^{\frac{1}{k}} \right| & \text{for } n = 2k, \\ \left| \pi - \left(\frac{n!\psi_n(d(\cdot))}{k!2^n n^n} \right)^{\frac{1}{k}} \right| & \text{for } n = 2k+1 \end{cases} \qquad (4.6)$$

In the following subsection, expressions for $surf\,(d(\cdot); r)$ and $vol(d(\cdot); r)$ for various m-neighbor distances are discussed and compared against the Euclidean hyperspheres in the sense of the above error measures.

4.1.2 Hyperspheres of m-Neighbor Distance

In a digital space, computing surface and volume as defined above is equivalent to counting the number of integral points satisfying a set of equations. A few results [29], which are used in computing the above are reviewed below. These are obtained using binomial expansion and enumerations for solutions of simple integer equations.

Lemma 4.1 $\forall r, t \in \mathcal{N}$ *the following holds:* $h_n(r, s) =$

$$(-1)^n \sum_{i=\lceil \frac{s}{r} \rceil}^{n} \binom{n}{i} (-2)^i \times \left[\sum_{j=0}^{\lfloor \frac{s}{r+1} \rfloor} \binom{i}{j} \binom{s+i-1-(r+1)j}{i-1} (-1)^j \right]$$

where $h_n(r, s)$ = *the number of distinct ways to select* $\bar{x}$ *from* $\mathcal{Z}^n$ *to satisfy the equation*

$$\sum_{i=1}^{n} |x_i| = s, 0 \le x_i \le r, 1 \le i \le n$$

The above expression is a polynomial of degree $n - 1$ in s with rational coefficients.

There are a few special cases of $h_n(r, s)$ as stated in the following corollary:

Corollary 4.1

$$h_n(r, s) = \begin{cases} 0 & s < 0, \\ 1 & s = 0, \\ (-1)^n \sum_{t=1}^{n} \binom{n}{t} (-2)^t \binom{t-1+s}{s} & r \ge s. \end{cases} \qquad (4.7)$$

Using the above result, we could compute the surface area of a disk of $d_m^{(n)}$ of radius r. In this case, we have to count the number of distinct solutions of $\bar{x} \in \mathcal{Z}^n$ for $d_m^{(n)}(\bar{x}) = r$. Thus, we get the following theorem [29]:

Theorem 4.1 $\forall m, n \in \mathcal{N}$, $m \le n$, and $r \in \mathcal{N}$, $surf\,(m, n; r) =$

$$\sum_{k=1}^{\lfloor \frac{m(r-1)}{r} \rfloor} \left[\sum_{s=0}^{(m-k)r-m} h_{n-k}(r-1, s) \right] \binom{n}{k} .2^k + \sum_{s=m(r-1)+1}^{mr} h_n(r, s)$$

where $h_n(r, s)$ is as defined in Lemma 4.1.

Theorem 4.2 $vol(m, n; r) = \sum_{i=0}^{r} surf\,(m, n; s)$

The above two theorems show that $surf\,(m, n; r)$ and $vol(m, n; r)$ are polynomials in r with rational coefficients of degree $n - 1$ and n, respectively. As $vol(m, n; r)$ is a polynomial of degree n in r, in [29], its coefficients are obtained by solving a set of $n + 1$ simultaneous equations, which are formed from the values of volume for $n + 1$ distinct values of $r = 0, 1, 2, \ldots, n$.

A few other properties that could be derived from the above theorems are described in the following [29]:

Corollary 4.2 *The following hold for surfaces and volumes:*

1. $\mathcal{H}(d_m^{(n)}; r) \subset \mathcal{H}(d_m^{(n)}; r + 1)$ and $vol(m, n; r) < vol(m, n; r + 1)$.
2. $\mathcal{S}(d_m^{(n)}; r) \cap \mathcal{S}(d_m^{(n)}; r + 1) = \phi$ but $surf\,(m, n; r) < surf\,(m, n; r + 1)$.
3. $\mathcal{H}(d_m^{(n)}; r) \subset \mathcal{H}(d_{m+1}^{(n)}; r)$ and $vol(m, n; r) < vol(m + 1, n; r)$.
4. $surf\,(m, n; r) < surf\,(m + 1, n; r)$ but $\mathcal{S}(d_m^{(n)}; r)$ and $\mathcal{S}(d_{m+1}^{(n)}; r)$ are unrelated.

From the above theorem, expressions of surfaces and volumes of hyperspheres of $d_m^{(n)}$ of radius r in $\mathcal{Z}^n$, $2 \le n \le 5$, are shown in Table 4.2 from [29]. We may note that as the underlying distance measure $d_m^{(n)}$ maintains an order, these polynomials preserve order over m and n. Corresponding expressions of surfaces and volumes of Euclidean hyperspheres are also shown in the table.

As $r \to \infty$, the coefficients of the highest degree term in the polynomials dominate the expression. Let this coefficient for the volume polynomial for the distance $d_m^{(n)}$ be denoted as $v_{m,n}$. Then the volume error $(E_V^{(n)}(d_m^{(n)}))$ of this distance is given by $\mid v_{m,n} - L_n \mid$. Further we observe that the coefficient corresponding to the highest degree term of the surface polynomial of the distance $d_m(n)$ (say, $s_{m,n}$) is related to that of the volume expression such that $s_{m,n} = n v_{m,n}$. Hence, the surface and shape errors are scaled values of the volume error, such that, $E_S^{(n)}(d_m^{(n)}) = n E_V^n(d_m^{(n)})$, and $E_\psi^{(n)}(d_m^{(n)}) = n^n E_V^{(n)}(d_m^{(n)})$. As these latter errors are inflated with the increase of dimensions, and also they could be obtained from the volume error, we provide only the volume error for a distance function in Table 4.2. We may note that for all the m-neighbor distances enumerated in the table in dimension n, the minimum volume error occurs at $m = 2$ (highlighted in the bold font).

Table 4.2 Volume and surface polynomials [29] for hyperspheres ($n = 2, 3, 4$, and 5)

n	m	$surf(d_m^{(n)})$	$vol(d_m^{(n)})$	$S_n^E(r)$	$V_n^E(r)$	$E_V^{(n)}(d(\cdot))$
2	1	$4r$	$1 + 2r + 2r^2$	$2\pi r$	πr^2	1.416
2	2	$8r$	$1 + 4r + 4r^2$			**0.8584**
3	1	$2 + 4r^2$	$1 + \frac{8}{3}r + 2r^2 + \frac{4}{3}r^3$	$4\pi r^2$	$\frac{4}{3}\pi r^3$	2.8555
3	2	$2 - 4r + 20r^2$	$1 + \frac{10}{3}r + 8r^2 + \frac{20}{3}r^3$			**2.4779**
3	3	$2 + 24r^2$	$1 + 6r + 12r^2 + 8r^3$			3.812
4	1	$\frac{16}{3}r + \frac{8}{3}r^2$	$1 + \frac{8}{3}r + \frac{10}{3}r^2 + \frac{4}{3}r^3 + \frac{2}{3}r^4$	$2\pi^2 r^3$	$\frac{1}{2}\pi^2 r^4$	4.2681
4	2	$16r - 16r^2 + 32r^3$	$1 + \frac{16}{3}r + 8r^2 + \frac{32}{3}r^3 + 8r^4$			**3.0652**
4	3	$\frac{32}{3}r - 8r^2 + \frac{184}{3}r^3$	$1 + 4r + \frac{50}{3}r^2 + 28r^3 + \frac{4}{3}r^4$			10.3985
4	4	$16r + 64r^3$	$1 + 8r + 24r^2 + 32r^3 + 16r^4$			11.6652
5	1	$2 + \frac{20}{3}r^2 + \frac{4}{3}r^4$	$1 + \frac{46}{15}r + \frac{10}{3}r^2$ $+ \frac{8}{3}r^3 + \frac{2}{3}r^4 + \frac{4}{15}r^5$	$\frac{8}{3}\pi^2 r^4$	$\frac{8}{15}\pi^2 r^5$	4.9971
5	2	$2 - \frac{32}{3}r + 52r^2$ $- \frac{88}{3}r^3 + 36r^4$	$1 + \frac{62}{15}r + \frac{40}{3}r^2$ $+ \frac{44}{3}r^3 + \frac{32}{3}r^4 + \frac{36}{5}r^5$			**1.9362**
5	3	$2 + \frac{7}{3}r + 60r^2$ $- \frac{176}{3}r^3 + 124r^4$	$1 + \frac{46}{5}r + \frac{50}{5}r^2 + 32r^3 +$ $\frac{142}{5}r^4 + \frac{124}{5}r^5$			19.5362
5	4	$2 - 8r + \frac{196}{3}r^2$ $- 8r^3 + \frac{476}{3}r^4$	$1 + \frac{18}{5}r + \frac{80}{3}r^2$ $+ \frac{212}{3}r^3 + \frac{232}{15}r^4 + \frac{476}{15}r^5$			26.4695
5	5	$2 + 80r^2 + 160r^4$	$1 + 10r + 40r^2$ $+ 80r^3 + 80r^4 + 32r^5$			26.7362

4.1.2.1 Vertices of Hyperspheres

In Table 4.1, the vertices of the hyperspheres of $d_m^{(n)}$ of radius r in 2-D and 3-D are enumerated. In the following theorem, we state their generalization to any arbitrary dimension:

Theorem 4.3 *The vertices of a hypersphere of radius r defined by the distance $d_m^{(n)}$ are given by $\phi(\bar{x})$ where*

$$\bar{x} = (\underbrace{r, r, \ldots, r}_{m}, \underbrace{0, 0, \ldots, 0}_{n-m})$$

and $\phi(\cdot)$ is the 2^n Symmetry Function (Definition 4.5).

4.1.2.2 Hyperspheres of Real m-Neighbor Distance

The definition of a hypersphere $\mathcal{H}(d_m^{(n)}; r)$ of $d_m^{(n)}$ in $\mathcal{Z}^n$ can easily be extended in $\mathcal{R}^n$ using the *real m-neighbor distance* $\delta_m^{(n)}$. For this distance function in the continuous space, we can likewise obtain expressions of its volume, surface, and vertices.

Definition 4.10 An *m-hypersphere* $\mathcal{H}(\delta_m^{(n)}; r)$ in *n*-D, of radius r and center $\bar{0}$, is defined as a subset of $\mathcal{R}^n$ as follows:

$$\mathscr{H}(\delta_m^{(n)}; r) = \{\bar{x} : \bar{x} \in \mathscr{R}^n, \ \delta_m^{(n)}(\bar{x}) \le r\} \tag{4.8}$$

As $\mathscr{H}$ is 2^n-symmetric, while deriving the properties, we may restrict our computation in the all positive hyperoctant $\mathscr{R}^{+^n}$ where $x_i \ge 0, \forall i, 1 \le i \le n$.

Definition 4.11 The *volume* $|\mathscr{H}(\delta_m^{(n)}; r)|$ of a hypersphere of $\delta_m^{(n)}$ is defined as
$$|\mathscr{H}(\delta_m^{(n)}; r)| =$$

$$\int\limits_{\delta_m^{(n)}(\bar{x}) \le r} d\bar{x}, \quad \text{where } d\bar{x} = dx_1 dx_2 \ldots dx_n \ \text{ and } \ \bar{x} = (x_1, x_2, \ldots, x_n). \tag{4.9}$$

In [21], the expression for $|\mathscr{H}(\delta_m^{(n)}; r)|$ has been computed using the inclusion–exclusion principle from combinations. We state the results in the following theorem:

Theorem 4.4 $\forall n \in \mathcal{N}, 0 < m \le n, r \ge 0$, *the volume of the hypersphere* $\mathscr{H}(\delta_m^{(n)}; r)$ *of* $\delta_m^{(n)}$ *is given by* $|\mathscr{H}(\delta_m^{(n)}; r)| = v_n(m) \cdot r^n$, *where*

$$v_n(m) = \frac{2^n}{n!} \sum_{j=0}^{\lceil m \rceil - 1} \binom{n}{j} \cdot (m-j)^n \cdot (-1)^j \tag{4.10}$$

In [21], the vertices of the hypersphere are also derived as stated below.

Theorem 4.5 *The hypersphere of* $\delta_m^{(n)}$ *of radius* r $(\mathscr{H}(\delta_m^{(n)}; r))$ *is a polytope whose vertices are* $\phi(\bar{x})$, *where*

$$\bar{x} = (\underbrace{r, r, \ldots, r}_{\lfloor m \rfloor}, (m - \lfloor m \rfloor)r, \underbrace{0, 0, \ldots, 0}_{n - \lceil m \rceil}) \tag{4.11}$$

and $\phi(\cdot)$ *is the* 2^n *symmetry function of n-D point as defined in Definition 4.5.*

From Theorems 4.3 and 4.5, we observe the following:

Corollary 4.3 *For integral m, hyperspheres of* $\delta_m^{(n)}$ *in* $\mathscr{R}^n$ *and* $d_m^{(n)}$ *in* $\mathscr{Z}^n$ *of the same radius have exactly the same set of vertices.*

There are also interesting results reported in [21], on the inscribed and circumscribed Euclidean hyperspheres. The Euclidean hyperspheres that can be *inscribed within* a hypersphere of $\delta_m^{(n)}$ is termed *insphere* and its radius is called *inradius*. We denote the inradius of a hypersphere of $\delta_m^{(n)}$ r_I. Likewise, the *circumscribed* Euclidean hypersphere is called *circumsphere*, whose radius, *circumradius*, is denoted by r_C. Precise mathematical definition of these radii are given in the following equations:

$$\begin{aligned} r_I &= \max\{\rho : \mathscr{H}(E^{(n)}; \rho) \subseteq \mathscr{H}(\delta_m^{(n)}; r)\} \ and \\ r_C &= \min\{\rho : \mathscr{H}(E^{(n)}; \rho) \supseteq \mathscr{H}(\delta_m^{(n)}; r)\} \end{aligned} \tag{4.12}$$

where $\mathcal{H}(E^{(n)}; \rho)$ is an Euclidean hypersphere of radius ρ with the center at origin. We may note that r_I and r_C both are functions of m, n, and r.

From the above definitions, it can be shown that $\mathcal{H}(E^{(n)}; r_I)$ and $\mathcal{H}(E^{(n)}; r_C)$ touch the $\mathcal{H}(\delta_m^{(n)}; r)$ at the furthest inner points $\bar{\iota}_I$ and nearest outer points $\bar{\iota}_C$, where $\delta_m^{(n)}(\bar{\iota}_I) = \delta_m^{(n)}(\bar{\iota}_C) = r$ and $E^{(n)}(\bar{\iota}_I) = r_I$ and $E^{(n)}(\bar{\iota}_C) = r_C$. The following theorem states the expressions of these quantities as functions of m, n, and r [21].

Theorem 4.6 $\forall n, n \geq 1, 0 < m \leq n$, we have

1. $r_I = \min\left(1, \frac{m}{\sqrt{n}}\right) \cdot r,$

2. $r_C = \sqrt{(\lfloor m \rfloor + (m - \lfloor m \rfloor)^2)} \cdot r,$ and
   $$\bar{\iota}_I \in \phi(r, 0, 0, \ldots, 0) \quad \text{for } m \leq \sqrt{n}$$
   $$\in \phi\left(\frac{mr}{n}, \frac{mr}{n}, \ldots, \frac{mr}{n}\right) \quad \text{for } m \geq \sqrt{n} \text{ and}$$

3.
   $$\bar{\iota}_C \in \phi(\underbrace{r, r, \ldots, r}_{\lfloor m \rfloor}, (m - \lfloor m \rfloor)r, \underbrace{0, 0, \ldots, 0}_{n - \lceil m \rceil})$$

where $\phi(\cdot)$ is the 2^n symmetry function of an n-D point.

4.2 Hyperspheres of Hyperoctagonal Distances

In this section the properties of the hyperspheres of HODs defined by a neighborhood sequence B are explored.

4.2.1 Vertices of Hyperspheres and Approximations

For a *well-behaved* neighborhood sequence, the vertices of a $\mathcal{H}(d_B^{(n)}; r)$ are given by the following theorem from [19].

Theorem 4.7 *For a well-behaved neighborhood sequence B in n-D, a hypersphere of the distance function $d_B^{(n)}$ has its vertices at $\phi(\bar{x})$ (where $\phi(\cdot)$ is the 2^n symmetry function in Definition 4.5), with $\bar{x}$ computed as follows:*

$$x_i = \left\lfloor \frac{r}{p} \right\rfloor \cdot (f_i(p) - f_{i-1}(p)) + f_i(r \bmod p) - f_{i-1}(r \bmod p), 1 \leq i \leq n$$

where

$$
\begin{aligned}
&\textit{Neighborhood sequence}: && B = \{b(1), b(2), \ldots, b(p)\} \\
&\textit{Length}: && p = |B| \\
&\textit{Trimmed } B: && B_i = \{b_i(1), b_i(2), \ldots, b_i(p)\}, \\
& && b_i(j) = \min(b(j), i), \ \forall i, 1 \le j \le p \\
&\textit{Sum sequence}: && F_i = \{f_i(1), f_i(2), \ldots, f_i(p)\}, \\
& && f_i(j) = \sum_{k=1}^{j} b_i(k), \forall i, 1 \le i \le p \ and \\
& && F_0 = \{0, 0, \ldots, 0\} \ and \ f_i(0) = 0
\end{aligned}
\tag{4.13}
$$

For $r < p$ some of the vertices may merge to form degenerate circles and spheres in both 2-D and 3-D. The following bounds hold for all neighborhood sequences:

Corollary 4.4 $\forall n, B$ and r, $x_1 = r$, and $0 \le x_i \le r$, $\forall i, 1 \le i \le n$.

Let us consider a few examples in 2-D and 3-D for illustration.

Example 4.1 Let $n = 2, B = \{1, 2, 1, 2, 2\}, p = |B| = 5$ and $r = 7$. Hence, the vertices of the hypersphere of $d^{(2)}_{12122}$ with radius 7 are given as

$$
B_2 = \{1, 2, 1, 2, 2\}, B_1 = \{1, 1, 1, 1, 1\} \ and
$$
$$
F_2 = \{1, 3, 4, 6, 8\}, F_1 = \{1, 2, 3, 4, 5\}
$$

$$
\begin{aligned}
x_1 &= \lfloor \tfrac{r}{5} \rfloor \cdot (f_1(5) - f_0(5)) + f_1(r \bmod 5) - f_0(r \bmod 5) & (4.14)\\
&= \lfloor \tfrac{7}{5} \rfloor \cdot (5 - 0) + (7 \bmod 5 - 0) & = 7 \\
x_2 &= \lfloor \tfrac{r}{5} \rfloor \cdot (f_2(5) - f_1(5)) + f_2(r \bmod 5) - f_1(r \bmod 5) \\
&= \lfloor \tfrac{7}{5} \rfloor \cdot (8 - 5) + (3 - 7 \bmod 5) & = 4
\end{aligned}
$$

With $r = 7$, the vertices are $\{(\pm 7, \pm 4), (\pm 4, \pm 7)\}$.

Example 4.2 Let $n = 3, B = \{1, 1, 2\}, p = |B| = 3$ and $r = 4$. Hence, the vertices of the hypersphere are given as

$$
B_3 = \{1, 1, 2\}, B_2 = \{1, 1, 2\}, B_1 = \{1, 1, 1\} \ and
$$
$$
F_3 = \{1, 2, 4\}, F_2 = \{1, 2, 4\}, F_1 = \{1, 2, 3\}
$$

$$
\begin{aligned}
x_1 &= \lfloor \tfrac{r}{3} \rfloor \cdot (f_1(3) - f_0(3)) + f_1(r \bmod 3) - f_0(r \bmod 3) \\
&= \lfloor \tfrac{4}{3} \rfloor \cdot (3 - 0) + (4 \bmod 3 - 0) & = 4 & \quad (4.15)\\
x_2 &= \lfloor \tfrac{r}{3} \rfloor \cdot (f_2(3) - f_1(3)) + f_2(r \bmod 3) - f_1(r \bmod 3) \\
&= \lfloor \tfrac{4}{3} \rfloor \cdot (4 - 3) + (4 \bmod 3 - 4 \bmod 3) & = 1 \\
x_3 &= \lfloor \tfrac{r}{3} \rfloor \cdot (f_3(3) - f_2(3)) + f_3(r \bmod 3) - f_2(r \bmod 3) \\
&= \lfloor \tfrac{4}{3} \rfloor \cdot (4 - 4) + (1 - 1) & = 0
\end{aligned}
$$

With $r = 4$, the vertices are given by

$$
\{(\pm 4, \pm 1, 0), (\pm 1, \pm 4, 0), (0, \pm 1, \pm 4), (0, \pm 4, \pm 1), (\pm 4, 0, \pm 1), (\pm 1, 0, \pm 4)\}
$$

In 2-D, the expression for vertices as given in Theorem 4.7, is further simplified as follows:

$$x_1 = r \text{ and } x_2 = \left\lfloor \frac{r}{p} \right\rfloor \cdot (f(p) - p) + f(r \bmod p) - (r \bmod p)$$

Further, as HODs generalize m-neighbor distances in n-D by taking the neighborhood sequence $B = \{m\}$ with $|B| = p = 1$, the vertices of its hyperspheres are also obtained from the expression of Theorem 4.7. These vertices are the same as given in Theorem 4.3, for $d_m^{(n)}$ (over $\mathscr{Z}^n$), i.e., 2^n-symmetry of the point $\bar{x} \in \mathscr{Z}^n$ as given below.

$$\bar{x} = (\underbrace{r, r, \ldots, r}_{m}, \underbrace{0, 0, \ldots, 0}_{n-m})$$

4.2.2 Hyperspheres of Sorted Neighborhood Sequences in 2-D and 3-D

We discuss here simplified expressions for computing vertices of hyperspheres for distances with sorted neighborhood sequences in 2-D and 3-D. A sorted neighborhood sequence in n-D is also conveniently represented by a fixed vector of dimension n, say Ω, whose ith component ω_i denotes the number of times the *type-i* neighborhood occurs in the sequence B. For example, in 3-D, $B = \{1, 1, 1, 3, 3\}$ can be equivalently represented by $\Omega = [3, 0, 2]$. With this representation, the length of the period of the sequence can be computed as $|B| = p = \sum_{i=1}^{n} \omega_i$. Simplified expressions of vertices with approximations are available in 2-D and 3-D for disks and spheres of HODs with intrinsically sorted neighborhood sequences [47]. Typical examples of disks and spheres of these distances are shown in Figs. 4.1 and 4.2. We may note that these distances satisfy the properties of a metric [23]. In Theorem 4.8 [47] and Theorem 4.9 [47], we state the approximations of vertices of disks and spheres in 2-D and 3-D, respectively. These approximations provide their representation in the real space, $\mathscr{R}^2$, and $\mathscr{R}^3$, respectively. These are used to derive a set of geometric measures related to the disks, such as perimeter, area, and shape feature for a 2-D disk, and surface area, volume, and shape feature for a 3-D sphere.

Following the fixed vector representation, in 2-D, we represent B as a doublet, $B = [\omega_1, \omega_2] = \{1^{\omega_1}, 2^{\omega_2}\}$ and $\omega_1 + \omega_2 = |B| = p$. In our notation, $\{1^m\}$ denotes a sequence of 1 of length m. We also denote a vertex of a disk of radius r by $\bar{x}(r) \in \mathscr{Z}^2$, and its approximation by $\bar{\bar{x}}(r) \in \mathscr{R}^2$, where $\bar{\bar{x}}(r) = \left(r, \frac{r\omega_2}{p}\right)$. Then it can be shown that the difference between $\bar{x}(r)$ and $\bar{\bar{x}}(r)$ is bounded as stated in the following theorem [47]:

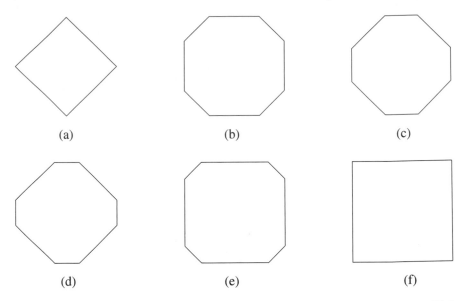

Fig. 4.1 Digital circles of 2-D octagonal distances for sorted neighborhood sequences: **a** {1}, **b** {1, 2}, **c** {1, 1, 2}, **d** {1, 1, 1, 2}, **e** {1, 2, 2}, and **f** {2}

Theorem 4.8 *For any $r > 0$ and for any sorted B, if its vertex in the all positive octant is given by $\bar{x}(r) = (x_1(r), x_2(r)) \in \mathscr{L}^2$, so that $x_1(r) \geq x_2(r)$ then*

1. $r - x_1(r) = 0$;

2. $0 \leq \left\{ \frac{r\omega_2}{p} - x_2(r) \leq \omega_1 - \frac{\omega_1^2}{p} \right\} \leq \begin{cases} \frac{p}{4} & p \text{ even} \\ \frac{(p^2 - 1)}{4p} & p \text{ odd} \end{cases}$

Similarly, in 3-D, a sorted B is represented as a triplet, $B = [\omega_1, \omega_2, \omega_3] = \{1^{\omega_1}, 2^{\omega_2}, 3^{\omega_3}\}$ and $\omega_1 + \omega_2 + \omega_3 = | B | = p$. The approximation of a vertex $\bar{x}(r) \in \mathscr{L}^3$, is given by $\tilde{\bar{x}}(r)$, where $\tilde{\bar{x}}(r) = \left(r, \frac{r(\omega_2 + \omega_3)}{p}, \frac{r\omega_3}{p} \right)$. Then the difference between $\bar{x}(r)$ and $\tilde{\bar{x}}(r)$ is bounded by the following theorem from [47]:

Theorem 4.9 *For any $r > 0$ and for any sorted B, if its vertex in the all positive hyperoctant is given by $\bar{x}(r) = (x_1(r), x_2(r), x_3(r)) \in \mathscr{L}^3$, so that $x_1(r) \geq x_2(r) \geq x_3(r)$, then*

1. $r - x_1(r) = 0$;

2. $0 \leq \begin{cases} \frac{r(\omega_2 + \omega_3)}{p} - x_2(r) \leq \omega_1 - \frac{\omega_1^2}{p} \\ \frac{r\omega_3}{p} - x_3(r) \leq \omega_3 - \frac{\omega_3^2}{p} \end{cases} \leq \begin{cases} \frac{p}{4} & p \text{ even} \\ \frac{(p^2 - 1)}{4p} & p \text{ odd} \end{cases}$

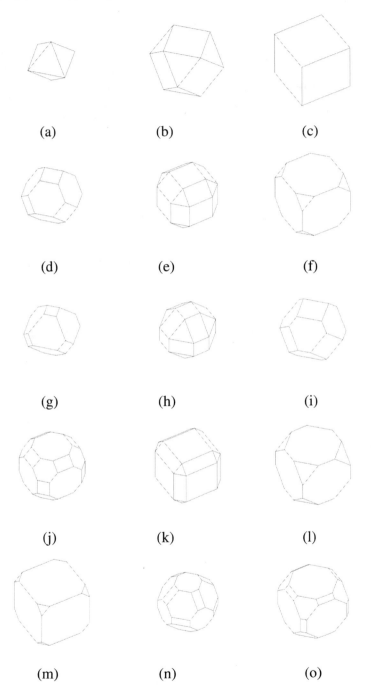

(a) (b) (c)

(d) (e) (f)

(g) (h) (i)

(j) (k) (l)

(m) (n) (o)

Fig. 4.2 Digital spheres of 3-D HODs for sorted neighborhood sequences: **a** {1}, **b** {2}, **c** {3}, **d** {1, 2}, **e** {1, 3}, **f** {2, 3}, **g** {1, 1, 2}, **h** {1, 1, 3}, i {1, 2, 2}, **j** {1, 2, 3}, **k** {1, 3, 3}, **l** {2, 2, 3}, **m** {2, 3, 3}, **n** {1, 1, 1, 2, 2, 3}, and **o** {1, 2, 2, 2, 3, 3}

4.2.2.1 Perimeter, Area, Volume, and Shape Features

From the coordinates of vertices, by applying the conventional treatment of coordinate geometry it is possible to derive different geometric measures of these disks (circles[1] and spheres) as functions of radius. From [38], we state a few theorems related to the computation of the geometric features of the digital disks (and spheres) of distances with intrinsically sorted neighborhood sequences.

Theorem 4.10 *The perimeter P and area A of a digital disk of radius r for an octagonal distance defined by a neighborhood sequence* $B = \{1^{\omega_1} 2^{\omega_2}\}$ *in 2-D are given by*

$$surf(\{1^{\omega_1}, 2^{\omega_2}\}; r) = P(\beta; r) = 4rP(\beta) \text{ and}$$
$$vol(\{1^{\omega_1}, 2^{\omega_2}\}; r) = A(\beta; r) = r^2 F(\beta) \tag{4.16}$$

where $P(\beta) = (2 - \sqrt{2})\beta + \sqrt{2}$, $F(\beta) = 2 + 4\beta - 2\beta^2$ *and* $\beta = \frac{\omega_2}{\omega_1 + \omega_2} = \frac{\omega_2}{p}$

Theorem 4.11 *The volume V and the surface area A of the polyhedron of radius r for a 3-D hyperoctagonal metric defined by a neighborhood sequence* $B = \{1^{\omega_1} 2^{\omega_2} 3^{\omega_3}\}$ *are given by*

$$surf(\{1^{\omega_1}, 2^{\omega_2}, 3^{\omega_3}\}; r) = A(\beta; r) = 4r^2 G(\beta, \gamma) \text{ and}$$
$$vol(\{1^{\omega_1}, 2^{\omega_2}, 3^{\omega_3}\}; r) = V(\beta; r) = \tfrac{4}{3} r^3 T(\beta, \gamma) \tag{4.17}$$

In the above
$T(\beta, \gamma) = 1 + 3\beta + 3\gamma + 6\beta\gamma + 3\beta^2 - 6\gamma^2 + 3\beta\gamma^2 - 6\beta^2\gamma - 2\beta^3 + \gamma^3$,
$G(\beta, \gamma) = \beta^2(3 - 2\sqrt{3}) + \gamma^2(\sqrt{3} - 3) + \beta\gamma(2\sqrt{3} - 6\sqrt{2} + 6) + \beta(2\sqrt{3}) + \gamma(6\sqrt{2} - 4\sqrt{3}) + \sqrt{3}$,
$\beta = \frac{\omega_2 + \omega_3}{\omega_1 + \omega_2 + \omega_3} = \frac{\omega_2 + \omega_3}{p}$, *and* $\gamma = \frac{\omega_3}{\omega_1 + \omega_2 + \omega_3} = \frac{\omega_3}{p}$.

For the digital sphere corresponding to the distance for which both β and γ are 0 in the above theorem, an octagonal face degenerates to a point, a rectangular one to a straight line segment and a hexagon to a triangle. For $\beta \neq 0$, and $\gamma = 0$, an octagon degenerates to a square, a rectangle to a straight line and a hexagon to a triangle. When values of both β and γ are 1, the octagon degenerates to a square, the rectangle to a straight line, and the hexagon to a point. In all such cases, the expressions given in Theorem 4.11 hold.

From the above theorems, the shape features in 2-D and 3-D are computed. In [47], the definition of the shape feature in 2-D is slightly different from the definitions used in this book (refer to Definition 4.4). The expressions of shape features with the definitions are given in Eqs. (4.18) and (4.19), respectively.

$$\psi_2(\beta) = \frac{(\text{perimeter})^2}{(\text{area})}$$
$$= \frac{(4rP(\beta))^2}{r^2 F(\beta)} \tag{4.18}$$
$$= \frac{16(\beta(2 - \sqrt{2}) + \sqrt{2})^2}{2 + 4\beta - 2\beta^2}$$

---

[1] In [34], it is discussed how different measures related to an Euclidean circle can be closely approximated by an octagon.

where $\beta = \frac{\omega_2}{\omega_1+\omega_2} = \frac{\omega_2}{p}$, and $P(\cdot)$ and $F(\cdot)$ are defined in Theorem 4.10.

$$\begin{aligned}
\psi_3(\beta, \gamma) &= \frac{\text{area}^3}{\text{volume}^2} \\
&= \frac{(4r^2 G(\beta,\gamma))^3}{(\frac{4}{3}r^3 T(\beta,\gamma))^2} \\
&= \frac{36(G(\beta,\gamma))^3}{(T(\beta,\gamma))^2}
\end{aligned} \tag{4.19}$$

where $\beta = \frac{\omega_2+\omega_3}{\omega_1+\omega_2+\omega_3} = \frac{\omega_2+\omega_3}{p}$, $\gamma = \frac{\omega_3}{\omega_1+\omega_2+\omega_3} = \frac{\omega_3}{p}$ and $G(\cdot, \cdot)$ and $T(\cdot, \cdot)$ are defined in Theorem 4.11.

4.2.3 Closeness to Euclidean Metric

From the above theorems on various measurements of disks and spheres of HODs defined by sorted neighborhood sequences in 2-D and 3-D, we compute geometric errors such as π-errors (refer to Definition 4.9) to measure the proximity to Euclidean metrics. These are defined below in the present context.

π-error measures in 2-D:

$$\text{PerimeterError} : E_{p\pi}^{(2)} = |\, \pi - 2P(\beta)\,|$$

$$\text{AreaError} : E_{a\pi}^{(2)} = |\, \pi - F(\beta)\,| \text{ and}$$

$$\text{Shapefeatureerror} : E_{\psi\pi}^{(2)} = \left|\, \pi - \frac{4(\beta(2-\sqrt{2})+\sqrt{2})^2}{2+4\beta-2\beta^2}\,\right|$$

$$= |\, \pi - S(\beta)\,|$$

where $S(\beta) = \frac{4(\beta(2-\sqrt{2})+\sqrt{2})^2}{2+4\beta-2\beta^2}$

π-error measures in 3-D:

$$\text{Volumetricerror} : E_{v\pi}^{(3)} = |\, \pi - T(\beta, \gamma)\,|$$

$$\text{Surfaceareaerror} : E_{s\pi}^{(3)} = |\, \pi - (G(\beta, \gamma))\,|$$

$$\text{Shapefeatureerror} : E_{\psi\pi}^{(3)} = \left|\, \pi - \frac{G(\beta, \gamma)^3}{T(\beta, \gamma)^2}\,\right|$$

By minimizing the above errors, it is possible to obtain optimum values of β and γ in 3-D, and that of β in 2-D. In 2-D, for minimizing the error measures, we need to solve the following equations separately:

Fig. 4.3 Plots of $E_{p\pi}^{(2)}$, $E_{a\pi}^{(2)}$, and $E_{\psi\pi}^{(2)}$ versus β

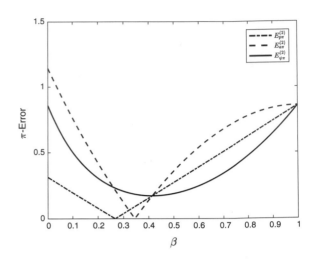

1. For perimeter error

$$P(\beta) = \pi/2, \quad 0 \le \beta \le 1 \tag{4.20}$$

2. For area error

$$F(\beta) = \pi, \quad 0 \le \beta \le 1 \tag{4.21}$$

3. For shape feature error

$$S(\beta) = \pi, \quad 0 \le \beta \le 1 \tag{4.22}$$

By solving Eq. (4.20), we get the value of β as $\frac{\pi - 2\sqrt{2}}{4 - 2\sqrt{2}} \approx 0.2673$ where $E_{p\pi}^{(2)}$ becomes zero. Likewise, the solution of Eq. (4.21), provides $E_{a\pi}^{(2)}$ as zero at $\beta = \frac{4 - \sqrt{32 - 8\pi}}{4} \approx$ 0.3449. The minimum value of $E_{\psi\pi}^{(2)}$ occurs at $\beta = \sqrt{2} - 1 \approx 0.4142$. It is interesting to note that at this value all types of π-errors have the same value, which is $8(\sqrt{2} - 1) \approx 0.1721$. In Fig. 4.3, plots of π-errors with varying β are shown. Theoretically, it is possible to set the value of β as close as possible over a large neighborhood sequence of length p (say $p \geq 1000$). But a large neighborhood sequence will make the shape of a digital circle of radius less than the length p nonuniform. For images of small size, this would also be nonrealistic. Hence, it is desirable to keep the length of neighborhood sequence small. Considering this, let us observe the properties of octagonal distances with the neighborhood sequence of length not more than 4 in 2-D. In Table 4.3, π-errors of octagonal distances with sorted neighborhood sequences till length 4 are enumerated. The minimum errors are highlighted in the table. We find that the closest distance for $p \leq 4$ is with the neighborhood sequence $B = \{1, 1, 1, 2\}$ which has minimum $E_{p\pi}^{(2)}$ among the candidates. For $E_{a\pi}^{(2)}$, the minimum value occurs

Table 4.3 Performances of a few 2-D octagonal distances defined by sorted neighborhood sequences using π-error measures

B	$E_{a\pi}^{(2)}$	$E_{p\pi}^{(2)}$	$E_{\psi\pi}^{(2)}$
{1}	1.142	0.313	0.858
{12}	0.358	0.273	0.189
{112}	**0.030**	0.078	0.189
{1112}	0.267	**0.020**	0.247
{122}	0.636	0.468	0.307
{1222}	0.733	0.566	0.405
{11122}	0.138	0.155	**0.173**
{2}	0.858	0.858	0.858

with the distance defined by the neighborhood sequence $B = \{1, 1, 2\}$. However, a close realization of optimum β providing minimum $E_{\psi\pi}^{(2)}$ is the octagonal distance with the neighborhood sequence $B = \{1, 1, 1, 2, 2\}$. This distance has $\beta = 0.4$ with $E_{\psi\pi}^{(2)} \approx 0.173$ (Table 4.3).

In 3-D, at different combinations of β and γ minimum error occurs for each error measure. In this case, the following equations are to be solved separately:

1. For surface-π error

$$G(\beta, \gamma) = \pi, \quad 0 \le \beta \le 1, 0 \le \gamma \le \beta \tag{4.23}$$

2. For volume-π error

$$T(\beta, \gamma) = \pi, \quad 0 \le \beta \le 1, 0 \le \gamma \le \beta \tag{4.24}$$

3. For shape-π error

$$\frac{(G(\beta, \gamma))^3}{(T(\beta, \gamma)^2} = \pi, \quad 0 \le \beta \le 1, 0 \le \gamma \le \beta \tag{4.25}$$

We observe that the domain of solution for the surface-π error lies within a region bounded by $0.2940 \le \beta \le 0.4310$ and $0.0014 \le \gamma \le 0.2925$.

Similarly, we compute optimum values of β and γ by minimizing $E_{v\pi}^{(3)}$. In [47], to carry out this analysis, for the ease of computation, an approximation of Eq. (4.24), has been considered without considering the γ^3 term. This is given below.

$$\gamma^2(3\beta - 6) + \gamma(-6\beta^2 + 6\gamma + 3) + (-2\beta^3 + 3\beta^2 + 3\beta + 1) = \pi \quad (4.26)$$

For the above equation, the domain of solution is the region bounded $0.3330 \le \beta \le 0.5310$ and $0.0005 \le \gamma \le 0.3286$. In Table 4.4, π-errors are enumerated for a few HODs in 3-D, whose neighborhood sequences are sorted. Out of them, we find that

Table 4.4 Performances of a few 3-D HODs defined by sorted neighborhood sequences using π-error measures

B	$E_{v\pi}^{(3)}$	$E_{a\pi}^{(3)}$	$E_{\psi\pi}^{(3)}$
{1}	2.142	1.410	2.050
{2}	1.854	1.590	1.090
{3}	2.858	2.858	2.858
{12}	0.142	0.2064	1.028
{13}	1.108	0.913	0.548
{23}	2.733	2.541	2.176
{112}	0.882	0.3064	1.323
{113}	**0.044**	**0.1804**	**0.470**
{122}	0.599	0.694	0.890
{123}	1.636	1.289	0.670
{133}	2.007	1.603	0.889
{223}	2.562	2.295	1.797
{233}	2.821	2.718	2.515
{11123}	0.1144	0.241	0.510
{1111223}	0.083	0.246	0.598

the distance with $B = \{1, 1, 3\}$ has the least errors for all the three measures. It should also be noted that for this distance both the values of β and γ are 0.33, the closest realization of a point in the solution space. There are also a few good distances having low values in one of these errors, such as {1111223}, and {11123}.

4.3 Analysis Using Properties of Hyperspheres in Real Space

In the geometric methods, a more pragmatic approach is to look for properties of hyperspheres of the distance functions in a real space instead of an integral space. It makes the analysis simpler in a continuous domain. The results obtained in the real space are very close to those of digital hyperspheres. As the radius of the hypersphere gets larger, these deviations get smaller, and in the limiting case of radius tending to infinity, it makes the results in both $\mathscr{R}^n$ and $\mathscr{L}^n$ identical. In subsequent sections, we extend our analysis in $\mathscr{R}^n$. Nevertheless, we would also discuss the properties of digital hyperspheres, as and when applicable.

In Chap. 2, we discuss about a hierarchy of the distance functions (refer to Fig. 2.4). Primarily there are two types of distance functions which establish these hierarchies, namely chamfering weighted distances (CWD) and weighted t-cost distances (WtD). Both classes generalize a few known classes of distance functions. For example, m-neighbor and t-cost distances are special cases of WtD. Even a subclass of HODs can be approximately described by the same functional form of a WtD. On the other hand, CWD also includes t-cost distances. Hence, the properties of hyperspheres of CWDs could also be extended to those of t-cost distance functions, while m-neighbor and some of the HODs can take advantage of the representational form of a WtD to assume its properties. That is why in the following sections, we discuss properties of hyperspheres of distances of these two broad classes and extend them to their special cases.

4.4 Hyperspheres of Weighted Distances

The vertices of hyperspheres of a weighted distance, represented in the form of an LWD, are given by the following theorem:

Theorem 4.12 *Let* $\gamma = \{\gamma_t, 1 \leq t \leq n\}$ *be the ordered set of weights for the* LWD *such that* $\gamma_1 \geq \gamma_2 \geq \ldots \geq \gamma_n \geq 0$, *and* $\gamma_1 > 0$. *Then, the vertices of the hypersphere of radius* r *whose center is at origin, are given as follows*

$$\phi\left(\left(\frac{r}{\gamma_1}, \underbrace{0, \ldots, 0}_{n-1\ zeroes}\right)\right), \phi\left(\left(\frac{r}{\gamma_1+\gamma_2}, \frac{r}{\gamma_1+\gamma_2}, \underbrace{0, \ldots, 0}_{n-2\ zeroes}\right)\right), \ldots,$$

$$\phi\left(\left(\underbrace{\frac{r}{\sum\limits_{i=1}^{k}\gamma_i}, \ldots, \frac{r}{\sum\limits_{i=1}^{k}\gamma_i}}, \underbrace{0, \ldots, 0}_{n-k\ zeroes}\right)\right), \ldots, \phi\left(\left(\frac{r}{\sum\limits_{i=1}^{n}\gamma_i}, \frac{r}{\sum\limits_{i=1}^{n}\gamma_i}, \ldots, \frac{r}{\sum\limits_{i=1}^{n}\gamma_i}, \ldots, \frac{r}{\sum\limits_{i=1}^{n}\gamma_i}\right)\right).$$

Proof Without loss of generality, we consider a vertex in the form of $\bar{u} = (u_1, u_2, \ldots, u_n)$, such that $u_1 \geq u_2 \geq \ldots \geq u_n \geq 0$.

At a vertex $\bar{u}$, a value at one of its dimension, reaches at maximum such that $LWD(\bar{u}; \Gamma) = r$. Let us define a type-k vertex, as the one where the magnitude reaches maximum at kth dimension. We look for the maximum possible value of kth maxima from the set of coordinate values subject to the constraint that the distance of the vertex should not go beyond r. This is true when $u_1 = u_2 = \ldots = u_k$ and the magnitudes at the rest coordinate positions are zeroes, i.e., $u_i = 0$, for $k < i \leq n$.

Hence, using $LWD(\bar{u}; \Gamma) = r$, we get the vertex as $\left(\underbrace{\frac{r}{\sum\limits_{i=1}^{k}\gamma_i}, \ldots, \frac{r}{\sum\limits_{i=1}^{k}\gamma_i}}, \underbrace{0, \ldots, 0}_{n-k\ zeroes}\right).$

All the type-k vertices are signed permutation of the above solution of $\bar{u}$, for $1 \leq k \leq n$.

Hence the theorem. $\square$

Applying the transformation of LWD to CWD (refer to Eq. (2.5)), we get the following result:

Corollary 4.5 *For the $CWD^{(n)}(\bar{u}; \Delta)$ with the ordered set of weights Δ, the vertices are given by the following:*

$$\phi\left(\left(\underbrace{\frac{r}{\delta_1}, 0, 0, \ldots, 0}_{n-1\ zeroes}\right)\right), \phi\left(\left(\frac{r}{\delta_2}, \frac{r}{\delta_2}, \underbrace{0, 0, \ldots, 0}_{n-2\ zeroes}\right)\right), \ldots, \phi\left(\left(\frac{r}{\delta_k}, \ldots, \frac{r}{\delta_k}, \underbrace{0, \ldots, 0}_{n-k\ zeroes}\right)\right), \ldots,$$
$$\phi\left(\left(\frac{r}{\delta_n}, \frac{r}{\delta_n}, \ldots, \frac{r}{\delta_n}, \ldots, \frac{r}{\delta_n}\right)\right).$$

In [35], a *normalized polytope* is constructed from a chamfering mask, so that the vertices of this polytope are formed by normalizing the vectors in the mask with their corresponding weights. If we consider the normalized polytope of $CWD^{(n)}$ with its chamfering mask consisting of vectors formed by m-neighbors only, we get the vertices of a hypersphere of unit radius as given in Corollary 4.5. In [61], similar concept is introduced in 3D and the normalized polytope is named *equivalent rational ball* (refer to Definition 2.9).

Corollary 4.6 *The vertices of a hypersphere of $CWD_{eu}^{(n)}$, whose ordered set of weights is given by $\Delta = \{\sqrt{i} | 1 \le i \le n\}$ lie on the surface of Euclidean hypersphere of the same radius.*

From the above corollary, it is observed that the hypersphere of $CWD_{eu}^{(n)}$ is enclosed by an Euclidean hypersphere of the same radius. This also implies that, for every point $\bar{u}$ in the space except at the vertices of a hypersphere, Euclidean norm ($E^{(n)}(\bar{u})$) is less than $CWD_{eu}^{(n)}(\bar{u})$. At a vertex of a hypersphere of $CWD_{eu}^{(n)}(\bar{u})$, the distances are equal. We may also note that these points in the space lie on a few well-defined directions, which are given by the vectors drawn from the origin to a vertex of any radius, given by Theorem 4.12. The number of these directions in n-D is $3^n - 1$, which could be obtained from the following corollary:

Corollary 4.7 *The total number of vertices according to Theorem 4.12, is $3^n - 1$.*

Proof The total number is given by the following expression:

$$\sum_{i=1}^{n} 2^i \binom{n}{i} = (1 + 2)^n - 1 \tag{4.27}$$
$$= 3^n - 1$$

$\square$

We should note here that for hyperspheres of some CWDs, some of these vertices may be degenerated to become a part of the smooth edges or surfaces of the boundary. In Fig. 4.4, we show a few examples of circles and spheres of different types of CWDs in 2-D and 3-D. To demonstrate the property of the circumscribed Euclidean circle around a circle of $CWD_{eu}^{(2)}$, an Euclidean circle of the same radius is also drawn enclosing that of $CWD_{eu}^{(2)}$ in Fig. 4.4a.

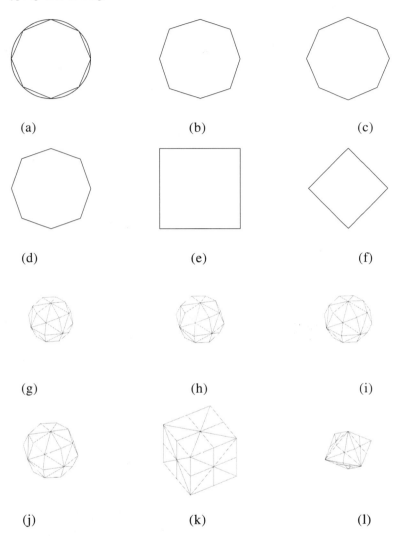

Fig. 4.4 Circles and spheres of: **a** $CWD_{eu}^{(2)}$, **b** $< 3, 4 >$, **c** $CWD_{opt}^{(2)}$, **d** $CWD_{opt*}^{(2)}$, **e** $< 1, 1 >$, **f** $< 1, 2 >$, **g** $CWD_{eu}^{(3)}$, **h** $< 3, 4, 5 >$, **i** $CWD_{opt}^{(3)}$, **j** $CWD_{opt*}^{(3)}$, **k** $< 1, 1, 1 >$, and **l** $< 1, 2, 3 >$

4.4.1 Volume and Surface

From the coordinates of vertices, the volume of a hypersphere is computed as given in the following theorem [43]:

Theorem 4.13 *The volume* $vol(CWD^{(n)}(\cdot; \Delta); r)$ *of an n-dimensional hypersphere of radius r for the* $CWD^{(n)}(\bar{u}; \Delta)$ *with the ordered set of weights* $\Delta(= \{\delta_1, \ldots, \delta_n\})$ *is given by the following:*

$$vol(CWD^{(n)}(\cdot; \Delta); r) = \frac{2^n}{\prod\limits_{i=1}^{n} \delta_i} r^n \tag{4.28}$$

Proof Let us consider the volume of a simplex wedge formed at origin with n vertices such as $(\frac{r}{\delta_1}, 0, 0, \ldots, 0)$, $(\frac{r}{\delta_2}, \frac{r}{\delta_2}, 0, \ldots, 0)$, $\ldots$, and, $(\frac{r}{\delta_n}, \frac{r}{\delta_n}, \ldots, \frac{r}{\delta_n})$, in the positive subspace . The volume V_s of this simplex is given by the following:

$$V_s = \frac{1}{n!} \begin{vmatrix} \frac{r}{\delta_1} & 0 & 0 & \ldots & 0 \\ \frac{r}{\delta_2} & \frac{r}{\delta_2} & 0 & \ldots & 0 \\ \frac{r}{\delta_3} & \frac{r}{\delta_3} & \frac{r}{\delta_3} & \ldots & 0 \\ \vdots & \vdots & \vdots & \vdots & 0 \\ \frac{r}{\delta_n} & \frac{r}{\delta_n} & \frac{r}{\delta_n} & \cdots & \frac{r}{\delta_n} \end{vmatrix}$$

$$= \frac{1}{n! \prod\limits_{i=1}^{n} \delta_i} r^n \tag{4.29}$$

As the hypersphere is partitioned into $2^n n!$ number of similar simplexes, the total volume of the hypersphere is obtained as $\frac{2^n}{\prod\limits_{i=1}^{n} \delta_i} r^n$. Hence, the theorem. $\square$

The surface area of a hypersphere is computed by the following theorem [43].

Theorem 4.14 *The surface area surf* $(CWD^{(n)}(\cdot; \Delta); r)$ *of an n-dimensional hypersphere of radius r for the* $CWD^{(n)}(\bar{u}; \Delta)$ *with the ordered set of weights as* $\Delta = (\{\delta_1, \delta_2, \ldots, \delta_n\})$ *is given by the following:*

$$surf(CWD^{(n)}(\cdot; \Delta); r) = n \frac{2^n}{\prod\limits_{i=1}^{n} \delta_i} f(\delta_1, \delta_2, \ldots, \delta_n) r^{n-1} \tag{4.30}$$

where

$$f(\delta_1, \delta_2, \ldots, \delta_n) = \sqrt{\delta_1^2 + \sum_{i=2}^{n} (\delta_i - \delta_{i-1})^2} \tag{4.31}$$

Proof For computing the area of the hypersurface formed by n vertices such as $(\frac{r}{\delta_1}, 0, 0, \ldots, 0)$, $(\frac{r}{\delta_2}, \frac{r}{\delta_2}, 0, \ldots, 0)$, ..., and, $(\frac{r}{\delta_n}, \frac{r}{\delta_n}, \ldots, \frac{r}{\delta_n})$, the surface normal, $\vec{\mu} = \alpha_1 \vec{j_1} + \alpha_2 \vec{j_2} + \ldots + \alpha_n \vec{j_n}$, is computed from the following expression, where $\vec{j_n}$, $\vec{j_2}$, ..., and $\vec{j_n}$, are unit vectors along the directions of the coordinate axes.

$$\vec{\mu} = \alpha_1 \vec{j_1} + \alpha_2 \vec{j_2} + \ldots + \alpha_n \vec{j_n}$$

$$= \begin{vmatrix} \vec{j_1} & \vec{j_2} & \vec{j_3} & \cdots & \vec{j_n} \\ \frac{r}{\delta_2} - \frac{r}{\delta_1} & \frac{r}{\delta_2} & 0 & \cdots & 0 \\ \frac{r}{\delta_3} - \frac{r}{\delta_1} & \frac{r}{\delta_3} & \frac{r}{\delta_3} & \cdots & 0 \\ \vdots & \vdots & \vdots & \vdots & 0 \\ \frac{r}{\delta_n} - \frac{r}{\delta_1} & \frac{r}{\delta_n} & \frac{r}{\delta_n} & \cdots & \frac{r}{\delta_n} \end{vmatrix}$$

(4.32)

The right hand side determinant expression is further simplified in the following form by subtracting ith row from $(i + 1)$th row, for $i = n, n - 1, \ldots, 3$, i.e.

$$\vec{\mu} = \begin{vmatrix} \vec{j_1} & \vec{j_2} & \vec{j_3} & \cdots \vec{j_{n-1}} & \vec{j_n} \\ \frac{r}{\delta_2} - \frac{r}{\delta_1} & \frac{r}{\delta_2} & 0 & \cdots 0 & 0 \\ \frac{r}{\delta_3} - \frac{r}{\delta_2} & \frac{r}{\delta_3} - \frac{r}{\delta_2} & \frac{r}{\delta_3} & \cdots 0 & 0 \\ \vdots & \vdots & \vdots & \vdots 0 & 0 \\ \frac{r}{\delta_{n-1}} - \frac{r}{\delta_{n-2}} & \frac{r}{\delta_{n-1}} - \frac{r}{\delta_{n-2}} & \frac{r}{\delta_{-1}n} - \frac{r}{\delta_{n-2}} & \cdots \frac{r}{\delta_{n-1}} & 0 \\ \frac{r}{\delta_n} - \frac{r}{\delta_{n-1}} & \frac{r}{\delta_n} - \frac{r}{\delta_{n-1}} & \frac{r}{\delta_n} - \frac{r}{\delta_{n-1}} & \cdots \frac{r}{\delta_n} - \frac{r}{\delta_{n-1}} & \frac{r}{\delta_n} \end{vmatrix}$$

(4.33)

By collecting cofactors of the $\vec{j_i}$ s, we obtain the following expressions:

$$|\alpha_1| = \frac{1}{\delta_2 \delta_3 \ldots \delta_n} r^{n-1}$$

$$|\alpha_2| = \left(\frac{1}{\delta_1} - \frac{1}{\delta_2}\right) \frac{1}{\delta_3 \ldots \delta_n} r^{n-1}$$

$$|\alpha_3| = \frac{1}{\delta_1} \left(\frac{1}{\delta_2} - \frac{1}{\delta_3}\right) \frac{1}{\delta_4 \ldots \delta_n} r^{n-1}$$

$$\vdots$$ (4.34)

$$|\alpha_i| = \frac{1}{\delta_1 \delta_2 \delta_{i-2}} \left(\frac{1}{\delta_{i-1}} - \frac{1}{\delta_i}\right) \frac{1}{\delta_{i+1} \ldots \delta_n} r^{n-1}$$

$$\vdots$$

$$|\alpha_i| = \frac{1}{\delta_1 \delta_2 \ldots \delta_{n-2}} \left(\frac{1}{\delta_{n-1}} - \frac{1}{\delta_n}\right) r^{n-1}$$

From the magnitude of $\vec{\mu}$, the area S_s of the hypersurface of n points is obtained as follows:

$$S_s = \frac{1}{(n-1)!}\sqrt{\vec{\mu}\cdot\vec{\mu}}$$

$$= \frac{1}{(n-1)!}\sqrt{\sum_{i=1}^{n}\alpha_i^2}$$

$$= \frac{1}{(n-1)!}\frac{1}{\prod\limits_{i=1}^{n}\delta_i}\sqrt{\delta_1^2 + \sum_{i=2}^{n}(\delta_i - \delta_{i-1})^2 r^{n-1}} \qquad (4.35)$$

By multiplying S_s with $2^n n!$ number of similar hypersurfaces, total surface area is given as by Eq. (4.30). Hence, the the theorem.   □

Following Eq. (4.31), in our discussion $f(\delta_1, \delta_2, \ldots, \delta_n)$ denotes the expression $\sqrt{\delta_1^2 + \sum_{i=2}^{n}(\delta_i - \delta_{i-1})^2}$. We also denote $(f(.))^k$ in a short form as $f^k(.)$.

4.4.2 Hyperspheres of t-Cost Distances

A t-cost distance $D_t^{(n)}$ can be represented by a CWD with the set weights as $\Delta = \{1, 2, 3, \ldots, \underbrace{t, \ldots, t}_{(n-t+1)}\}$. From Corollary 4.5, we obtain the vertices of t-cost distances as stated in the following theorem:

Theorem 4.15 *The vertices of a hypersphere of radius r of the norm $D_t^{(n)}$ are given by*

$$\phi\left(\left(r, \underbrace{0, 0, \ldots, 0}_{n-1\ zeroes}\right)\right), \phi\left(\left(\frac{r}{2}, \frac{r}{2}, \underbrace{0, 0, \ldots, 0}_{n-2\ zeroes}\right)\right), \ldots, \phi\left(\left(\frac{r}{t}, \ldots, \frac{r}{t}, \frac{r}{t}, \ldots, \frac{r}{t}, \underbrace{0, \ldots, 0}_{n-t\ zeroes}\right)\right),$$

$$\phi\left(\left(\frac{r}{t}, \frac{r}{t}, \ldots, \frac{r}{t}, \underbrace{0, \ldots, 0}_{n-t-1\ zeroes}\right)\right), \ldots, \phi\left(\left(\frac{r}{t}, \frac{r}{t}, \ldots, \frac{r}{t}\right)\right).$$

Similarly, volume and surface area of a hypersphere of $D_t^{(n)}$ are obtained from Theorems 4.13 and 4.14.

Theorem 4.16 *The volume $vol(D_t^{(n)}; r)$ of an n-dimensional hypersphere of radius r of $D_t^{(n)}$ is given by the following:*

$$vol(D_t^{(n)}; r) = \frac{2^n}{t! t^{(n-t)}} r^n \qquad (4.36)$$

Theorem 4.17 *The surface area surf $(D_t^{(n)}; r)$ of an n-dimensional hypersphere of radius r of $D_t^{(n)}$ is given by the following:*

$$surf\,(D_t^{(n)}; r) = n\frac{\sqrt{t}2^n}{t!t^{(n-t)}}r^{n-1} \tag{4.37}$$

Proof $f(1, 2, \ldots, \underbrace{t, t, \ldots, t}_{(n-t+1)}) = \sqrt{t}$ (from Eq. (4.31)). From Eq. (4.30), we get the above expression. $\qquad\square$

4.4.3 Proximity to Euclidean Distance

For studying how close are the distances to Euclidean metric, different π-errors as defined in Definition 4.9, are used in [43]. They are reformulated using the properties of hyperspheres of $CWD^{(n)}(\bar{u}; \Delta)$, $< \delta_1, \delta_2, \ldots, \delta_n >$.

$$E_{s\pi}^{(n)}(CWD^{(n)}(\bar{u}; \Delta)) = \begin{cases} \left|\left|\pi - \left(\dfrac{2^{2k}f(\delta_1,\delta_2,\ldots,\delta_n)k!}{\prod\limits_{i=1}^{n}\delta_i}\right)^{\frac{1}{k}}\right|\right| & \text{for } n = 2k, \\[5em] \left|\left|\pi - \left(\dfrac{f(\delta_1,\delta_2,\ldots,\delta_n)(2k+1)!}{k!\prod\limits_{i=1}^{n}\delta_i}\right)^{\frac{1}{k}}\right|\right| & \text{for } n = 2k + 1. \end{cases} \tag{4.38}$$

$$E_{v\pi}^{(n)}(CWD^{(n)}(\bar{u}; \Delta)) = \begin{cases} \left|\left|\pi - \left(\dfrac{2^{2k}k!}{\prod\limits_{i=1}^{n}\delta_i}\right)^{\frac{1}{k}}\right|\right| & \text{for } n = 2k, \\[5em] \left|\left|\pi - \left(\dfrac{(2k+1)!}{k!\prod\limits_{i=1}^{n}\delta_i}\right)^{\frac{1}{k}}\right|\right| & \text{for } n = 2k + 1. \end{cases} \tag{4.39}$$

$$
E_{\psi\pi}^{(n)}(CWD^{(n)}(\bar{u}; \Delta)) = \begin{cases} \left| \left| \pi - \left(\dfrac{2^{2k} f^{2k}(\delta_1, \delta_2, \ldots, \delta_n) k!}{\displaystyle\prod_{i=1}^{n} \delta_i} \right)^{\frac{1}{k}} \right| \right| & \text{for } n = 2k, \\[3em] \left| \left| \pi - \left(\dfrac{f^{2k+1}(\delta_1, \delta_2, \ldots, \delta_n)(2k+1)!}{k! \displaystyle\prod_{i=1}^{n} \delta_i} \right)^{\frac{1}{k}} \right| \right| & \text{for } n = 2k + 1. \end{cases}
$$

$$(4.40)$$

The π-error measures for different CWDs in 2-D, 3-D, and 4-D are shown in Table 4.5. We may note that weights of distances $< 3, 4 >$, $< 3, 4, 5 >$, and $< 3, 4, 5, 6 >$ in 2-D, 3-D, and 4-D, respectively, are scaled by $\frac{1}{3}$ before computing the errors. In the table, the smallest error values among the distance functions are shown with the bold font. Empirically, we observe that $CWD_{opt}^{(2)}$, $CWD_{mse}^{(2)}$ and $CWD_{eu}^{(2)}$ in 2-D, $CWD_{opt}^{(3)}$, $CWD_{mse}^{(3)}$ and $CWD_{eu}^{(3)}$ in 3-D, and $CWD_{opt}^{(4)}$ and $CWD_{eu}^{(4)}$ in 4-D are good distances in this class for approximating Euclidean metrics with respect to these measures. In the previous chapter, we observe that these distances are also close to Euclidean metrics with low analytical and empirical errors. This shows that geometric error measures have good correlation with other analytical error measures to judge the quality of approximation of Euclidean metrics.

4.4.4 π-Errors of t-Cost Distances

By using Eqs. (4.36) and (4.37), π-errors of hyperspheres of t-cost distances are computed and optimum values of t in every dimension for each type of error are obtained. In Table 4.6, we present these values for every dimension. Empirically, it is observed that with increasing dimension optimum values of t slowly rise for surface and volume errors. Optimum shape-π errors are quite high compared to the other two optimum errors for surface and volume, and it remains roughly within a narrow range of 0.85 to 1.01. The optimum value of this error also takes place at a much higher value of t, which is roughly close to $\frac{n}{2}$ in n-D. We have observed a similar trend in a much higher dimensional space also. For example, at dimension 100 optimum values of t for surface, volume and shape π-errors take place at 5, 5, and 50, respectively.

Table 4.5 π-error measures of CWD in different dimensions

2-D	$E_{s\pi}^{(2)}$	$E_{v\pi}^{(2)}$	$E_{\psi\pi}^{(2)}$	3-D	$E_{s\pi}^{(3)}$	$E_{v\pi}^{(3)}$	$E_{\psi\pi}^{(3)}$	4-D	$E_{s\pi}^{(4)}$	$E_{v\pi}^{(4)}$	$E_{\psi\pi}^{(4)}$
$CWD_{euopt}^{(2)}$	0.046	0.075	**0.172**	$CWD_{euopt}^{(3)}$	**0.013**	0.191	0.375	$CWD_{euopt}^{(4)}$	**0.054**	**0.162**	0.294
$< 3, 4 >$	0.041	0.141	0.192	$< 3, 4, 5 >$	0.157	0.442	0.507	$< 3, 4, 5, 6 >$	0.258	0.458	0.436
$CWD_{opt}^{(2)}$	0.085	0.083	0.173	$CWD_{opt}^{(3)}$	0.044	0.229	0.361	$CWD_{opt}^{(4)}$	0.096	0.209	**0.269**
$CWD_{opt*}^{(2)}$	**0.007**	0.180	0.184	$CWD_{opt*}^{(3)}$	0.073	0.337	0.532	$CWD_{opt*}^{(4)}$	0.078	0.245	0.482
$CWD_{umse}^{(2)}$	0.088	**0.005**	**0.172**	$CWD_{umse}^{(3)}$	0.104	**0.011**	**0.298**				
$< 1, 1 >$	1.717	0.858	0.858	$< 1, 1, 1 >$	2.858	2.858	2.858	$< 1, 1, 1, 1 >$	2.515	2.515	2.515
$< 1, 2 >$	0.626	1.142	0.858	$< 1, 2, 3 >$	1.410	2.142	2.055	$< 1, 2, 3, 4 >$	1.509	1.987	1.477

Table 4.6 Optimum values of t giving minimum π-errors of surface, volume, and shape in the class of t-cost distance functions

Dimension	Surface-π error		Volume-π error		Shape-π error	
n	$t_{opt}^{(s)}$	$min.E_{s\pi}^{(s)}$	$t_{opt}^{(v)}$	$min.E_{v\pi}^{(v)}$	$t_{opt}^{(\psi)}$	$min.E_{\psi\pi}^{(sh)}$
2	2	0.313	1	0.858	1	0.858
3	2	1.020	2	1.642	2	1.101
4	2	0.763	2	1.142	3	0.858
5	2	0.839	2	1.205	3	1.020
6	2	0.572	2	0.852	3	0.858
7	2	0.494	2	0.783	4	0.980
8	2	0.271	2	0.509	4	0.858
9	2	0.118	2	0.369	5	0.955
10	2	0.066	2	0.149	6	0.858
11	2	0.266	2	0.038	6	0.939
12	2	0.419	2	0.219	6	0.858
13	2	0.653	2	0.440	7	0.928
14	2	0.780	2	0.590	7	0.858
15	2	1.041	2	0.839	8	0.919
16	3	0.979	2	0.963	9	0.858
17	3	0.950	3	1.095	9	0.912
18	3	0.826	3	0.963	9	0.858
19	3	0.784	3	0.924	10	0.907
20	3	0.670	3	0.802	11	0.858
21	3	0.618	3	0.753	11	0.902
22	3	0.512	3	0.640	12	0.858
23	3	0.450	3	0.582	12	0.899
24	3	0.352	3	0.477	12	0.858
25	3	0.283	3	0.411	13	0.896
26	3	0.192	3	0.314	13	0.858
27	3	0.115	3	0.240	14	0.893
28	3	0.030	3	0.150	14	0.858
29	3	0.053	3	0.070	15	0.891
30	3	0.132	3	0.014	15	0.858
31	3	0.221	3	0.100	16	0.889
32	3	0.295	3	0.179	16	0.858
33	3	0.389	3	0.269	17	0.887
34	3	0.458	3	0.343	18	0.858
35	3	0.557	3	0.439	18	0.885
36	3	0.621	3	0.508	18	0.858
37	3	0.724	3	0.608	19	0.884
38	3	0.785	3	0.673	20	0.858
39	4	0.785	3	0.777	20	0.883
40	4	0.722	4	0.805	21	0.858

4.4.4.1 Hypersphere of Distances Induced by Chamfer Mask of Larger Neighborhood

Vertices of hyperspheres of chamfer masks are computed directly from the vertices of *equivalent rational ball* (ERB) as discussed in Sect. 2.3.1. We summarize these observations in the following theorem:

Theorem 4.18 *Given the generator* $\mathscr{M}_g = \{(\bar{u}_i, w_i)|i = 1, 2, \ldots, M; \bar{u}_i \in \mathscr{Z}^n; w_i \in \mathscr{R}^+\}$ *of a chamfer mask, the vertices of its hypersphere of radius r of the induced norm are given by*

$$\mathscr{V} = \{\phi(\frac{r}{w_i}\bar{u}_i)|1 \le i \le M\}$$

The above statement is valid, if $\mathscr{V}$ forms a convex polytope.

Corollary 4.8 *Theorem 4.12 is a special case of Theorem 4.18.*

Example 4.3 Let us consider an example for a 5×5 chamfer mask in 2-D. Consider the generator of the mask given as $\{((1, 0), 5), ((1, 1), 7), ((2, 1), 11)\}$. Then the vertices of its circle of radius r are given by

$$\mathscr{V} = \left\{ \left(\pm\frac{r}{5}, 0\right), \left(0, \pm\frac{r}{5}\right), \left(\pm\frac{r}{7}, \pm\frac{r}{7}\right), \left(\pm\frac{2r}{11}, \pm\frac{r}{11}\right), \left(\pm\frac{r}{11}, \pm\frac{2r}{11}\right) \right\}$$

In Fig. 4.5a, the circle of the above CMID is shown. Similarly, we can compute the vertices in higher dimensions. In the same figure (Fig. 4.5b–d) a few typical examples in 3-D are shown.

Verwer [80] followed a simple geometric approach for computing the MRE of a CMID. In this case, he proposed to compute errors on the points lying on the

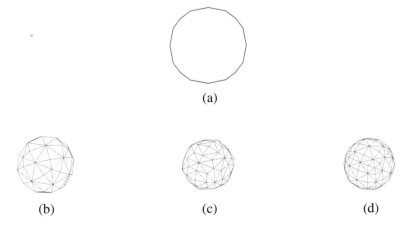

(a)

(b)   (c)   (d)

Fig. 4.5 **a** Circle of $5 - 7 - 11$ CMID of mask size 5×5. **b–d**: Spheres of chamfer masks of size $5 \times 5 \times 5$: **b** $a = 3, b = 4, c = 5, e = 7$, **c** $a = 8$. $b = 11, c = 14, d = 18, e = 20$, and **d** $a = 23$, $b = 32, c = 39, d = 51, e = 55, f = 68$

Table 4.7 MRE of CMIDs (including CWDs) in 2D from analysis by Verwer [80]

Distance	$\frac{1}{\kappa}$	MRE
$CWD_{eu}^{(2)}$	1.0412	0.0396
$< 2, 3 >$	2.1180	0.0557
$< 5, 7 >$	5.1675	0.0421
$< 12, 17 >$	12.5000	0.0400
$\mathcal{M}_{55}(4, 6, 9)$	4.1213	0.0294
$\mathcal{M}_{55}(5, 7, 11)$	5.0092	0.0179
$\mathcal{M}_{55}(9, 13, 20)$	9.0819	0.0152
$\mathcal{M}_{55}(17, 24, 38)$	17.2174	0.0143
$\mathcal{M}_{55}(1, \sqrt{2}, \sqrt{5})$	1.0137	0.0136

Euclidean circle of radius 1. The minimum and maximum chamfering distances of the points are obtained. Let these distances be denoted as d_{min} and d_{max}. Given a chamfering mask with the generator $\mathcal{M}_g = \{(\bar{u}_i, w_i)|i = 1, 2, \ldots, M; \bar{u}_i \in \mathcal{L}^n; w_i \in \mathcal{R}^+\}$, these are computed as given below.

$$d_{min} = \min_{i=1}^{M} \left\{ \frac{w_i}{||\bar{u}_i||} \right\}$$

where $||\bar{u}_i||$ is the magnitude of $\bar{u}_i$.

If the maximum angle between two neighboring vectors[2] is θ, then

$$d_{max} = \frac{1}{cos(\frac{\theta}{2})}$$

For example, for a mask of size $K \times K$, $\theta = tan^{-1}\frac{2}{K-1}$. Given d_{max} and d_{min} as computed above the MRE for a CMID with an optimum scale factor κ, is given by the following:

$$MRE = \frac{1 - cos(\frac{\theta}{2})}{1 + cos(\frac{\theta}{2})}, \text{ and } \kappa = \frac{1}{1 - MRE} \tag{4.41}$$

Verwer's approach is general enough to compute MRE with optimum scale factor for any arbitrary CMID in an arbitrary n-dimensional real space. He reported results in 2-D and 3-D. Intuitively, he also derived the $CWD_{euopt}^{(n)}$ as the distance with an optimal MRE, and extended it to any arbitrary mask, where w_i is assigned by the magnitude of the vector $\bar{u}$ in a mask. A few results from [80], in 2-D and 3-D are tabulated in Tables 4.7 and 4.8.

In a similar approach [12], exploiting the geometry of circles in 2-D, optimum CWD and CMIDs are reported. In this technique, the normalized deviation of points

---

[2]They form an angular section, which does not have any vector in the mask.

Table 4.8 MRE of CMIDs (including CWDs) in 3D from analysis by Verwer [80]

Distance	$\frac{1}{\kappa}$	MRE
$CWD_{eu}^{(3)}$	1.0641	0.0602
$< 2, 3, 4 >$	2.2247	0.1010
$< 3, 4, 5 >$	3.0725	0.0794
$< 4, 6, 7 >$	4.2913	0.0679
$< 7, 10, 12 >$	7.4011	0.0639
$< 19, 27, 33 >$	20.2355	0.0611
$\mathcal{M}_{555}(4, 6, 7, 9, 10, 12)$	4.1213	0.0294
$\mathcal{M}_{555}(9, 13, 16, 20, 22, 27)$	9.18913	0.0266
$\mathcal{M}_{555}(20, 29, 35, 45, 49, 60)$	20.5000	0.0244
$\mathcal{M}_{555}(1, \sqrt{2}, \sqrt{3}, \sqrt{5}, \sqrt{6}, \sqrt{9})$	1.0247	0.0241

in the circumference of a circle of unit radius of digital distances from that of the Euclidean circle of the same radius is expressed as a function of the angle formed at the center with the horizontal axis (x-axis). This provides a new optimization criterion compared to what has been used in [80]. In this approach, while defining the relative error, the normalization is carried out with respect to the chamfering distance. For example, given a point on the unit circle of the chamfering distance at an angle θ, if the Euclidean distance of the same point is $L(\theta)$, the relative error in this case[3] is defined as $| 1 - L(\theta) |$. Maximizing this error, we obtain a measure very close to our definition of MRE in Sect. 3.1. To distinguish it, let us call this measure as $MRE_{\mathcal{M}}$. The $MRE_{\mathcal{M}}$ of the distance function is expressed as the functions of weights. Further, minimization of the $MRE_{\mathcal{M}}$ in the parameter space is carried out. The technique has been applied for chamfering distances of varying mask sizes, such as 3×3, 5×5, and 7×7. For a distance function $CWD_{<a,b>}^{(2)}$ which is defined by a 3×3 chamfering mask, the optimum values of parameters are found to be at $a = 0.9619$ and $b = 1.3604$ providing an $MRE_{\mathcal{M}}$ of 0.0396. If the value of a is kept at 1, the optimum b is found to be 1.342 with the value of $MRE_{\mathcal{M}}$ as 0.0538. Some of the good approximations of these distances with integer weights are also reported as $\frac{1}{3} < 3, 4 >, (MRE_{\mathcal{M}} = 0.06066)$ and $\frac{1}{72.77} < 70, 99 > (MRE_{\mathcal{M}} = 0.02719)$. In a 5×5 mask the recommended distances with integer weights are $\frac{1}{5} < 5, 7, 11 > (MRE_{\mathcal{M}} = 0.01942)$, $\frac{1}{34.45} < 34, 48, 76 >$ $(MRE_{\mathcal{M}} = 0.1358)$, and $\frac{1}{30.57} < 30, 43, 67 > (MRE_{\mathcal{M}} = 0.02024)$.

Similar methodology has been adopted in defining mean squared errors of distances. The latter has been computed in three scenarios. They carried out unconstrained conventional mean squared error (MSE) minimization, and then constrained minimization by making the estimation unbiased, so that the average of error over the circumference becomes zero at the estimation point in the parameter space. The latter methodology is adopted from [80]. In the third approach, minimization of the

---

[3] According to Definition 3.3, it is otherwise $| 1 - \frac{1}{L(\theta)} |$.

MSE is carried out by keeping the areas of the circle of the chamfering distance and that of the Euclidean norm of the same radius the same. The distances which are found to provide low MREs are also reported to have low MSE values in constrained and unconstrained optimization problems.

4.4.5 Hyperspheres of Weighted t-Cost Distances

It is possible to compute vertices for a type of WtD, which has nonincreasing set of weights with increasing t as defined below.

Definition 4.12 A weight vector W is said to be *well-behaved* if the weights are positive numbers and ordered in nonincreasing manner. That is, $w_1 \geq w_2 \geq \cdots \geq w_n > 0$. The corresponding distance function is known as *well-behaved weighted t-cost distance* .

We call a weight vector is *strongly well-behaved* if the weights are strictly decreasing with increasing t and all of them are greater than 0, i.e $w_1 > w_2 > \ldots > w_n > 0$.

For a well-behaved W there exists a result for the vertices of a hypersphere in [42] (refer to Lemma 3 of the paper). However, that lemma has some limitations, as it does not apply the constraint on nonincreasing order of the coordinates of the vertices, which is assumed in the proposition. In [50], a modified version of the lemma is presented. We state it in the following theorem:

Theorem 4.19 *For a well-behaved W, vertices of a hypersphere of radius r are given by $\phi(\bar{u})$ where $\bar{u} \in \mathscr{R}^n$, $u_1 \geq u_2 \geq \ldots \geq u_n \geq 0$ and*

$$u_i = \begin{cases} \frac{r}{w_i} & \text{for } i = 1. \\ \min\left(\frac{r}{w_i} - \sum_{j=1}^{i-1} u_j, u_{i-1} \right) & \text{Otherwise.} \end{cases} \tag{4.42}$$

Proof As $u_1 \geq u_2 \geq \ldots \geq u_n \geq 0$, we try to maximize values of individual components by keeping the distance as r. In this case, the following conditions should hold:

$$w_1 D_1^n(\underline{u}) = w_2 D_2^n(\underline{u}) = w_3 D_3^n(\underline{u}) = \cdots = w_n D_n^n(\underline{u}) = r$$

So we get, $u_1 = \frac{r}{w_1}$.

Next, we need to solve, $w_2 D_2^n(\underline{u}) = r \Rightarrow w_2(u_1 + u_2) = r$.

By substituting the value of u_1, we get, $u_2 = (\frac{1}{w_2} - \frac{1}{w_1})r$.

In this way the ith equation is written as, $w_i(u_1 + u_2 + \cdots + u_i) = r$.

Assuming $w_{i-1} > 0$, we have two possible scenarios.

Case I: $w_i < w_{i-1}$, and $w_i \neq 0$

We can write the following by substituting the values of u_j's, $1 \leq j \leq (i-1)$.

$$u_i = \frac{r}{w_i} - \sum_{j=1}^{i-1} u_j = \left(\frac{1}{w_i} - \frac{1}{w_{i-1}} \right) r$$

Case II: $w_i = w_{i-1}$,

We get $u_i = 0$, and all subsequent coordinate values would be 0 for satisfying the constraint of nonincreasing order of nonnegative values.

Without loss of generality, the same would hold for all permutations of signed coordinate values. Hence the theorem.   □

From the above theorem, we observe the following:

Corollary 4.9 *Given a well-behaved W, if there exists j, such that $w_i = w_j, j+1 \leq i \leq n$, and $w_{j-1} > w_j$, the vertices are given by $\phi(\bar{u})$, where $\bar{u} \in \mathcal{R}^n$ and*

$$u_i = \begin{cases} \frac{r}{w_i} & for\ i = 1. \\ \left(\frac{1}{w_i} - \frac{1}{w_{i-1}} \right) r & 2 \leq i \leq j \\ 0 & j+1 \leq i \leq n \end{cases} \tag{4.43}$$

The result reported previously (Lemma 3 of [42]) is applicable for *strongly well-behaved* WtD. This is stated in the following corollary:

Corollary 4.10 *If $w_1 > w_2 > \ldots > w_n > 0$, the vertices of the hypersphere of radius r are given by $\phi(\bar{u})$ where $\bar{u} \in \mathcal{R}^n$ and*

$$u_i = \begin{cases} \frac{r}{w_i} & for\ i = 1. \\ \left(\frac{1}{w_i} - \frac{1}{w_{i-1}} \right) r & Otherwise. \end{cases} \tag{4.44}$$

The above corollary holds for *inverse square root weighted t-cost norm* $(WtD_{isr}^{(n)})$ whose weights are given by $w_t = \frac{1}{\sqrt{t}}$ for $1 \leq t \leq n$. Hence the vertices of the hypersphere of $WtD_{isr}^{(n)}$ are given by the following corollary:

Corollary 4.11 *The vertices of a hypersphere of $WtD_{isr}^{(n)}$ of radius r are $\phi(\bar{u})$, where $\bar{u} \in \mathcal{R}^n$ and*

$$u_i = \begin{cases} r & for\ i = 1. \\ (\sqrt{i} - \sqrt{i-1})r & Otherwise. \end{cases} \tag{4.45}$$

Examples of a digital circle and a sphere in 2-D and 3-D for inverse square root weighted *t*-cost norms are shown in Fig. 4.6a, b, respectively.

Fig. 4.6 **a** A digital circle
and **b** a digital sphere of
$WtD_{isr}^{(2)}$, and $WtD_{isr}^{(3)}$,
respectively

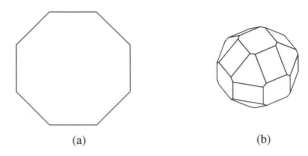

(a)   (b)

The shapes of circles and spheres for well-behaved WtDs are similar to those
of HODs in respective dimensions. Hence, using the expressions in Eq. (4.16), in
2-D and Eq. (4.17) in 3-D, different measures related to the shape of the circles
and spheres of *well-behaved weighted t-cost norm* in 2-D and 3-D are stated in the
following theorems. In these theorems, the ordered set of weights is normalized with
$w_1 = 1$.

Theorem 4.20 *The perimeter and area of the circle of radius r for a weighted t-cost
norm in 2-D with the ordered set of weights as* $\{1, w_2\}$, *such that* $0 < w_2 \leq 1$, *are*
$4rP(\beta)$ *and* $r^2 F(\beta)$, *where* $\beta = \frac{1}{w_2} - 1$, *and*

$$P(\beta) = (2 - \sqrt{2})\beta + \sqrt{2}, \tag{4.46}$$

and,

$$F(\beta) = 2 + 4\beta - 2\beta^2. \tag{4.47}$$

Theorem 4.21 *The surface area and volume of the sphere of radius r for a weighted
t-cost norm in 3-D with the ordered set of weights as* $\{1, w_2, w_3\}$, *such that* $0 < w_3 \leq w_2 \leq 1$, *are* $4r^2 G(\beta, \gamma)$ *and* $\frac{4}{3}r^3 T(\beta, \gamma)$, *where* $\beta = \frac{1}{w_2} - 1$, $\gamma = \frac{1}{w_3} - \frac{1}{w_2}$, *and*

$$G(\beta, \gamma) = (3 - 2\sqrt{3})\beta^2 + (\sqrt{3} - 3)\gamma^2 + (2\sqrt{3} - 6\sqrt{2} + 6)\beta\gamma + 2\sqrt{3}\beta + (6\sqrt{2} - 4\sqrt{3})\gamma + \sqrt{3}, \tag{4.48}$$

and,

$$T(\beta, \gamma) = 1 + 3\beta + 3\gamma + 6\beta\gamma + 3\beta^2 - 6\gamma^2 + 3\beta\gamma^2 - 6\beta^2\gamma - 2\beta^3 + \gamma^3. \tag{4.49}$$

From the above theorems, the measures of disks of inverse square root weighted
t-cost norms in 2-D and 3-D are obtained as stated in the following corollaries:

Corollary 4.12 *In 2-D, the perimeter and area of a circle of* $WtD_{isr}^{(2)}$ *of radius r are
given by* $6.6274r$ *and* $3.313708r^2$, *respectively. The circle is a regular octagon.*

Corollary 4.13 *In 3-D, the surface area and volume of* $WtD_{isr}^{(2)}$ *of radius r are given
by* $14.3319r^2$ *and* $4.7773r^3$, *respectively.*

Fig. 4.7 Circle of inverse square root weighted t-cost norm in 2-D, a regular octagon, enclosing the Euclidean circle of the same radius

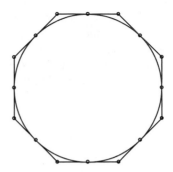

It is observed that for $WtD_{isr}^{(n)}$ in n-D, the hypersphere of radius r encloses the n-D Euclidean hypersphere of the same radius. Also, the Euclidean hypersphere touches all the center points of the faces (hyperplane of dimension $n-1$) of the hypersphere of $WtD_{isr}^{(n)}$. For example, in 2-D the circle of $WtD_{isr}^{(2)}$, which is a regular octagon of radius r, encloses the Euclidean circle of the same radius (refer to Fig. 4.7). The Euclidean circle touches each side of the regular octagon at its midpoints, that is, 2^n symmetry sets of vertices $(r, 0)$ and $(\frac{r}{\sqrt{2}}, \frac{r}{\sqrt{2}})$, $\phi((r, 0))$ and $\phi((\frac{r}{\sqrt{2}}, \frac{r}{\sqrt{2}}))$, respectively.

In the following theorem, we state the property of containment of the Euclidean hypersphere within that of inverse square root weighted t-cost norm of same radius [42].

Theorem 4.22 *The Euclidean hypersphere of radius r is enclosed within the hypersphere of $WtD_{isr}^{(n)}$ of the same radius. Also, the Euclidean hypersphere touches the hypersphere of $WtD_{isr}^{(n)}$ exactly at $3^n - 1$ points, which are centers of the hyperfaces (of dimension $n-1$) of the hypersphere.*

For example in 2-D and 3-D, the Euclidean circle and sphere touch the circle and sphere of $WtD_{isr}^{(2)}$ and $WtD_{isr}^{(3)}$ at 8 and 26 points, respectively. The points of contacts in 2-D are as mentioned before $\phi((r, 0))$ and $\phi((\frac{r}{\sqrt{2}}, \frac{r}{\sqrt{2}}))$. A typical example of this property is shown in Fig. 4.7. In 3-D, for a radius of r, these points are given by the sets $\phi((r, 0, 0))$, $\phi((\frac{r}{\sqrt{2}}, \frac{r}{\sqrt{2}}, 0))$, and $\phi((\frac{r}{\sqrt{3}}, \frac{r}{\sqrt{3}}, \frac{r}{\sqrt{3}}))$.

4.4.5.1 Hyperoctagonal Distances Approximated by WtD

In Sect. 2.5.3, we discuss a close approximation of a HOD $d_B^{(n)}$ in the form of a WtD. For the sake of continuity of discussion, let us review this form below.

$$\hat{d}_B^{(n)}(\bar{u}) = \max_{t=1}^{n} \left\{ \left\lceil \frac{pD_t^{(n)}(\bar{u})}{f_t(p)} \right\rceil \right\} \tag{4.50}$$

where $B = \{b(1), b(2), \ldots, b(p)\}$ is the neighborhood sequence of length $\mid B \mid = p$, and $f_t(p)$ is given as $\sum_{i=1}^{p} min(b(i), t)$, $1 \leq t \leq n$. As we are considering approximation of digital hyperspheres in a real space, we use the continuous form of approximation by ignoring the ceiling ($\lceil . \rceil$) operation. From the above, we represent a HOD by a WtD with the set of weights $W = \{w_t | w_t = \frac{p}{f_t(p)}\}$. Using this representation, we can compute vertices of hyperspheres of HODs, when W turns out to be well-behaved, and for a class of HODs, we could ascertain this property as stated in the following theorem:

Theorem 4.23 *Given a sorted neighborhood sequence B, with a fixed vector representation $\Omega = [\omega_1, \omega_2, \ldots, \omega_n]$ the weight vector $W = \{w_1, w_2, \ldots, w_n\}$ of its approximated WtD is well-behaved and normalized with the $w_1 = 1$.*

Proof For a sorted B, $f_t(p)$ is computed from components of Ω as follows:

$$f_t(p) = \begin{cases} p & \text{for } t = 1, \\ \sum_{i=1}^{t-1}(i.\omega_i) + (\sum_{i=t}^{n}\omega_i).t & 2 \leq t \leq n \end{cases} \tag{4.51}$$

Hence, $f_t(p)$ is monotonically increasing with t. This makes $w_t = \frac{p}{f_t(p)}$ monotonically decreasing. Again, as $f_1(p) = p$, $w_1 = 1$. Hence, the theorem. $\qquad\square$

As the set of weights is found to be well-behaved, we obtain the vertices of hyperspheres of approximated HODs with sorted neighborhood sequences as stated in the following theorem:

Theorem 4.24 *Given a sorted neighborhood sequence B with a fixed vector representation $\Omega = [\omega_1, \omega_2, \cdots, \omega_n]$ the vertices of the hypersphere of radius r of $\hat{d}_B^{(n)}$ are given by $\phi(\bar{u})$ where $\bar{u} \in \mathcal{R}^n$ and*

$$u_i = \begin{cases} r & \text{for } i = 1. \\ \alpha_i r & \text{Otherwise.} \end{cases} \tag{4.52}$$

where α_i is the fraction of neighborhood types greater or equal to the type i in the sequence B of length p and is given by

$$\alpha_i = \frac{\sum_{t=i}^{n}\omega_t}{p} \tag{4.53}$$

Proof From Eq. 4.51, we get the following:

$$\frac{1}{w_t} = 1 + \sum_{i=2}^{t}\alpha_i \tag{4.54}$$

Hence

$$\frac{1}{w_t} - \frac{1}{w_{t-1}} = \alpha_t, \text{ for } 2 \leq t \leq n \tag{4.55}$$

From Theorem 4.19, we get the expressions as given in the statement. $\qquad\square$

Corollary 4.14 *The vertices of a hypersphere of radius r of $d_m^{(n)}$ following the representation by a WtD are given by $\phi(\bar{u})$ where $\bar{u} \in \mathcal{R}^n$ and*

$$u_i = \begin{cases} r & \text{for } 1 \le i \le m \\ 0 & m < i \le n \end{cases} \tag{4.56}$$

Proof In the WtD form, weights W of $d_m^{(n)}$ are given by [42]

$$w_t = \begin{cases} \frac{1}{t}, & \text{for } t < m \\ \frac{1}{m}, & \text{for } t \ge m \end{cases} \tag{4.57}$$

From Corollary 4.9, we get the above result. $\qquad\square$

4.4.5.2 Proximity to Euclidean Hyperspheres

Following the approach of measuring the degree of proximity of HODs in 2-D and 3-D using π-errors, we compute the same for *well-behaved* WtDs in 2-D and 3-D. The error measures using the ordered set of weights are defined below.

Error measures in 2-D: For $w_1 = 1$, and $\beta = \frac{1}{w_2} - 1$,

$$\text{Perimeter} - \pi \text{Error} : E_{p\pi}^{(2)}(\beta) = |\pi - 2P(\beta)|,$$

$$\text{Area} - \pi \text{Error} : E_{a\pi}^{(2)}(\beta) = = |\pi - F(\beta)|, \text{ and}$$

$$\text{Shape} - \pi \text{Error} : E_{\psi\pi}^{(2)}(\beta) = \left| \pi - \frac{4(\beta(2 - \sqrt{2}) + \sqrt{2})^2}{2 + 4\beta - 2\beta^2} \right| = |\pi - S(\beta)|.$$

where $S(\beta) = \frac{4(\beta(2 - \sqrt{2}) + \sqrt{2})^2}{2 + 4\beta - 2\beta^2}$.

Error measures in 3-D: For $w_1 = 1$, $\beta = \frac{1}{w_2} - 1$, and $\gamma = \frac{1}{w_3} - \frac{1}{w_2}$,

$$\text{Surface} - \pi \text{ Error} : E_{s\pi}^{(3)}(\beta, \gamma) = |\pi - (G(\beta, \gamma))|,$$

$$\text{Volume} - \pi \text{ Error} : E_{v\pi}^{(3)}(\beta, \gamma) = |\pi - T(\beta, \gamma)|, \text{ and}$$

$$\text{Shape} - \pi \text{ Error} : E_{\psi\pi}^{(3)}(\beta, \gamma) = \left| \pi - \frac{G(\beta, \gamma)^3}{T(\beta, \gamma)^2} \right|.$$

In Theorems 4.20 and 4.21, the functional forms of $P(\beta)$, $F(\beta)$, $T(\beta, \gamma)$, and $G(\beta, \gamma)$ are provided. In Tables 4.9 and 4.10, values of corresponding π-error measures are shown for some of the representative octagonal distances approximated by weighted t-cost distance functions. In their approximate representation in the form of WtD also, they have the same shape of hyperspheres as those derived for sorted neighborhood sequences. Hence, the error measures are exactly the same of what are

Table 4.9 Geometric error measures in 2-D. Smallest error measures are highlighted in bold font

B	$E_{p\pi}^{(2)}$	$E_{a\pi}^{(2)}$	$E_{\psi\pi}^{(2)}$
$\{1, 2\}$	0.273	0.358	0.189
$\{1, 1, 2\}$	0.077	**0.030**	0.189
$\{1, 1, 1, 2\}$	**0.020**	0.267	0.247
$\{1, 2, 2\}$	0.468	0.636	0.307
$\{1, 2, 2, 2\}$	0.565	0.733	0.405
$WtD_{isr}^{(2)}$	0.172	0.172	**0.172**

Table 4.10 Geometric error measures in 3-D. Smallest error measures are highlighted in bold font

B	$E_{s\pi}^{(3)}$	$E_{v\pi}^{(3)}$	$E_{\psi\pi}^{(3)}$
$\{1, 2\}$	0.206	0.142	1.028
$\{1, 3\}$	0.912	1.108	0.548
$\{2, 3\}$	2.541	2.733	2.176
$\{1, 1, 3\}$	**0.180**	**0.044**	0.472
$\{2, 2, 3\}$	2.295	2.562	1.797
$\{1, 1, 1, 2, 3\}$	0.241	0.114	0.509
$\{1, 1, 1, 1, 2, 2, 3\}$	0.246	0.083	0.598
$WtD_{isr}^{(2)}$	0.441	0.441	**0.441**

reported in Tables 4.3 and 4.4. In Tables 4.9 and 4.10, corresponding error measures for *inverse square root weighted t-cost norm* $WtD_{isr}^{(2)}$ in 2-D and $WtD_{isr}^{(3)}$ are also shown, respectively . We observe that all the error measures have low values with this norm. In particular, in both tables, their *shape-π errors are the smallest among the distance functions chosen for comparison.*

4.5   Geometric Computation of MRE

In this section, we discuss a geometric approach [44], for computing MRE of a distance function in $\mathscr{R}^n$. The strategy is similar to what has been adopted in [80]. We first discuss how MREs could be obtained for a norm which either overestimates or underestimates an Euclidean norm in $\mathscr{R}^n$. The definitions of overestimated and underestimated norms are given below.

Definition 4.13 A distance function is called an *overestimated norm* (OEN), if for every point in the space, its norm value is greater or equal to the Euclidean norm at that point.

Definition 4.14 A distance function is *an underestimated norm* (UEN) if the norm value is less or equal to the corresponding Euclidean value at every point in the space.

We should note that above nomenclatures are with respect to Euclidean norms. For example, $d_1^{(n)}$ (also called L_1 norm) is an *overestimated norm*, and $d_n^{(n)}$ (also called L_∞ norm) an underestimated norm. Other interesting examples include $CWD_{eu}^{(n)}$ and $WtD_{isr}^{(n)}$, for overestimated and underestimated norms, respectively. They have better bounds on these estimates than those of L_1 and L_∞ norms. It is observed in [2] (refer to Sect. 3.5.4) a scaled $CWD_{eu}^{(n)}$ is optimally closer to Euclidean norm $E^{(n)}$ among all CWDs, where the scale κ is given by

$$\kappa = \frac{2}{1 + \sqrt{\sum_{i=1}^{n} \left(\sqrt{i} - \sqrt{i-1}\right)^2}} \tag{4.58}$$

We refer to $\kappa CWD_{eu}^{(n)}$ as $CWD_{euopt}^{(n)}$.

In the class of t-cost norms [31], $D_1^{(n)}$ and $D_n^{(n)}$ are the same $L_\infty^{(n)}$ and $L_1^{(n)}$ norms, respectively. But there are a few more distance functions in the same class, which are also OENs as stated in the following theorem [20]:

Theorem 4.25 *For* $\lceil \sqrt{n} \rceil \leq t \leq n$, $D_t^{(n)}$ *is an overestimate of Euclidean norm.*

The proof is available in [20]. In the next section, we provide an intuitive explanation from the geometry of their hyperspheres.

4.5.1 Characterization by Hyperspheres

The overestimated and underestimated norms can easily be characterized by observing the containment relationship between hyperspheres of these norms and the corresponding Euclidean spheres of the same radii. In the following theorems, we state these relationships. We omit the proofs as they are trivial.

Theorem 4.26 *A hypersphere of an overestimated norm is enclosed by an Euclidean hypersphere of the same radius.*

Theorem 4.27 *A hypersphere of an underestimated norm encloses an Euclidean hypersphere of the same radius.*

Theorem 4.28 *A hypersphere of radius R is that of an overestimated norm, if and only if* *the value of the Euclidean norm at its every vertex is less than or equal to R.*

From Theorem 4.28, we easily get an algorithm for deciding whether a norm is overestimated, by checking the vertices of hyperspheres. The Theorem 4.25 also follows from this property. However, for an underestimated norm this simple vertex-check is not sufficient. It needs additional computation for checking the containment of an Euclidean hypersphere of the same radius as stated in the following theorem.

Theorem 4.29 *For a hypersphere of radius R of an underestimated norm, all its vertices are at an Euclidean distance from its center greater than or equal to R. In that case, a norm is underestimated, if and only if the shortest Euclidean distance from the center of its hypersphere of radius R to its boundary surface ($(n-1)$ space) is greater than or equal to R.*

From Theorem 4.28, and the expressions of vertices discussed in this chapter, we find that L_1, $CWD_{eu}^{(n)}$, and $\{D_t^{(n)}|\lceil\sqrt{n}\,\rceil \leq t \leq n\}$ are overestimated norms. These observations are also corroborated with the results reported in [31, 43], respectively, which were obtained by adopting analytical approaches involving analysis of functional forms of these norms. The simplicity of this geometric characterization shows us its power over the conventional methods of functional analysis. This would be further revealed, when we adopt a similar approach in obtaining MREs of different norms in $\mathscr{R}^n$ as discussed in the next section.

4.5.2   Computation of MRE from Hyperspheres

From the shape of a hypersphere of an underestimated norm, it is possible to compute the theoretical value of its MRE [44].

Theorem 4.30 *When the norm ρ is an underestimation of the Euclidean norm, the maximum deviation of the norm value for all points lying on the boundary of the hypersphere takes place at the vertices which are maximally furthest in the sense of Euclidean norm value. Let this maximum deviation for a unit hypersphere of the norm be denoted as d_{max}. In that case, the MRE of the norm is given by the following:*

$$MRE(\rho) = 1 - \frac{1}{d_{max}} \tag{4.59}$$

From the above theorem and also the expressions of vertices of some of the underestimated norms, we obtain their MREs as stated in following corollaries:

Corollary 4.15 *The MRE of $WtD_{isr}^{(n)}$ is given by [42]*

$$MRE(WtD_{isr}^{(n)}) = 1 - \frac{1}{\eta} \tag{4.60}$$

where

$$\eta = \sqrt{\sum_{i=1}^{n}(\sqrt{i} - \sqrt{i-1})^2} \tag{4.61}$$

Corollary 4.16 *The MRE of L_∞ (or $d_n^{(n)}$) is given by*

$$MRE(L_\infty^{(n)}) = 1 - \frac{1}{\sqrt{n}} \tag{4.62}$$

For overestimated norm, a point lying on the boundary surface of its hypersphere is maximally deviated from the Euclidean hypersphere of the same radius, which is at minimal Euclidean distance from the center. As we are considering hyperspheres of norms the center of a hypersphere lies at the origin of the space. With this observation, let us consider the hyperspheres of CWDs. They are convex polytopes. We obtain the minimal Euclidean distance from the origin to its boundary surface by computing the perpendicular distance on a hyperplane in its surface from the origin. The perpendicular distance is computed by the length of the projector on the hyperplane from the center (origin) along the direction of its normal. The following theorem provides the shortest distance from the origin to a boundary hyperplane of a hypersphere of radius R of a CWD, as well as its MRE, if the CWD is an overestimated norm.

Theorem 4.31 *The shortest Euclidean distance (d_s) between the center and a point in the boundary surface of a hypersphere of the CWD $< \delta_1, \delta_2, \ldots, \delta_n >$ of radius R is given by*

$$d_s = \frac{R}{\sqrt{\delta_1^2 + \sum_{i=2}^{n} (\delta_i - \delta_{i-1})^2}} \tag{4.63}$$

Hence, if the norm is an overestimation of an Euclidean norm, its MRE is given by

$$MRE(< \delta_1, \delta_2, \ldots, \delta_n >) = \sqrt{\delta_1^2 + \sum_{i=2}^{n} (\delta_i - \delta_{i-1})^2} - 1 \tag{4.64}$$

Proof Let us consider the hypersurface of a hypersphere of radius R formed by n vertices such as $(\frac{R}{\delta_1}, 0, 0, \ldots, 0)$, $(\frac{R}{\delta_2}, \frac{R}{\delta_2}, 0, \ldots, 0)$, $\ldots$, and, $(\frac{R}{\delta_n}, \frac{R}{\delta_n}, \ldots, \frac{R}{\delta_n})$ (refer to Corollary 4.5). Let its surface normal be $\vec{\mu} = \alpha_1 \vec{j_1} + \alpha_2 \vec{j_2} + \ldots + \alpha_n \vec{j_n}$, where $\vec{j_n}, \vec{j_2}, \ldots,$ and $\vec{j_n}$, are unit vectors along the directions of the coordinate axes.

From Eq. (4.34), we restate the following results:

$$
\begin{aligned}
|\alpha_1| &= \frac{1}{\delta_2 \delta_3 \ldots \delta_n} R^{n-1} \\
|\alpha_2| &= \left(\frac{1}{\delta_1} - \frac{1}{\delta_2} \right) \frac{1}{\delta_3 \ldots \delta_n} R^{n-1} \\
&\vdots \\
|\alpha_n| &= \frac{1}{\delta_1 \delta_2 \ldots \delta_{n-2}} \left(\frac{1}{\delta_{n-1}} - \frac{1}{\delta_n} \right) R^{n-1}
\end{aligned}
\tag{4.65}
$$

Table 4.11 MREs of overestimated CWD norms

Norm	Weight vectors	MRE
$L_1^{(n)}$	$< 1, 2, 3, \ldots, n >$	$\sqrt{n} - 1$
$CWD_{eu}^{(n)}$	$< 1, \frac{1}{\sqrt{2}}, \frac{1}{\sqrt{3}}, \ldots, \frac{1}{\sqrt{n}}$	$n - 1$ (refer to Eq. (4.61))
$D_t^{(n)}$ (for, $\lceil \sqrt{n} \rceil \leq t \leq n$)	$< 1, 2, 3, \ldots, t, \ldots, t >$	$\sqrt{t} - 1$

Perpendicular distance (d_s) from the origin to the hyperplane is given by $\overrightarrow{OV} \cdot \frac{\vec{\mu}}{|\vec{\mu}|}$, where $\overrightarrow{OV}$ is the vector formed from origin to any vertex, say $(\frac{R}{\delta_1}, 0, 0, \ldots, 0)$. Hence,

$$
\begin{aligned}
d_s &= \frac{\alpha_1 \cdot \frac{R}{\delta_1}}{\sqrt{\vec{\mu} \cdot \vec{\mu}}} \\
&= \frac{R}{\sqrt{\delta_1^2 + \sum_{i=2}^{n}(\delta_i - \delta_{i-1})^2}}
\end{aligned}
\tag{4.66}
$$

Thus for an overestimated norm in n-D, the MRE is given by

$$
\begin{aligned}
\frac{R - d_s}{d_s} &= \frac{R}{d_s} - 1 \\
&= \sqrt{\delta_1^2 + \sum_{i=2}^{n}(\delta_i - \delta_{i-1})^2} - 1
\end{aligned}
\tag{4.67}
$$

$\square$

Corollary 4.17 *The shortest Euclidean distance (d_s) of a point in the boundary surface of a hypersphere of the LWD with $\Gamma = (\gamma_1, \gamma_2, \ldots, \gamma_n)$ of unit radius from the origin is given by*[4]

$$
d_s = \frac{1}{\sqrt{\sum_{i=1}^{n} \gamma_i^2}}
\tag{4.68}
$$

Using above theorems, MREs of overestimated CWD norms are given in Table 4.11.

If a norm is neither an underestimation, nor an overestimation of the corresponding Euclidean norm, both the above results are combined to provide its MRE as stated in the following theorem [44].

Theorem 4.32 *Let d_{max} be the maximum Euclidean distance of any vertex from the center of the hypersphere of a norm ρ of unit radius, and b_{min} be the shortest Euclidean distance between the center and its boundary surface. The MRE of the norm is given by the following:*

---

[4] In the LWD functional form, Γ can be treated as an outer normal vector of a facet of the hypersphere. Hence, $d_s = \frac{1}{E^{(n)}(\Gamma)}$.

$$MRE(\rho) = \max \left(\left| 1 - \frac{1}{d_{max}} \right|, \left| \frac{1}{b_{min}} - 1 \right| \right) \tag{4.69}$$

From the above theorem, we get the MRE of a CWD as stated in the following theorem:

Theorem 4.33

$$MRE(CWD^{(n)}(\bar{u}; \Delta)) = \max \left(\left| 1 - \frac{1}{d_{max}} \right|, \left| 1 - \frac{1}{b_{min}} \right| \right) \tag{4.70}$$

where

$$d_{max} = \sqrt{\max_{t=1}^{n} \left\{ \frac{t}{\delta_t^2} \right\}}, \tag{4.71}$$

and

$$b_{min} = \frac{1}{\sqrt{\delta_1^2 + \sum_{i=2}^{n} (\delta_i - \delta_{i-1})^2}}. \tag{4.72}$$

From Theorem 4.33, the MRE of a *t-cost* distance[5] is obtained as follows:

Theorem 4.34

$$MRE(D_t^{(n)}) = \begin{cases} 1 - \frac{1}{\sqrt{n}}, & t = 1 \\ \max(\sqrt{t} - 1, 1 - \frac{t}{\sqrt{n}}), & 1 < t < \sqrt{n} \\ \sqrt{t} - 1, & \sqrt{n} \le t \le n. \end{cases} \tag{4.73}$$

We observe that the middle row of Eq. (4.73), depicts the general form for any arbitrary value t, which is also true for $t = 1$, and $t = n$. In [31], it was derived using functional analysis. The above theorem provides more precise expressions.

4.5.3 Optimum Scale Factor and Scale Adjusted MRE

It is possible to bring a distance function ρ more closer to Euclidean norm by scaling it [2, 10, 14]. It implies that there exists a scale factor κ, so that $\kappa.\rho$ provides the minimum MRE among all positive scale factors. In the following [46], we provide the expression for optimum scale factor (κ_{opt}) at which the MRE becomes minimum for a distance function ρ. We also provide the scale adjusted optimum MRE value.

Theorem 4.35 *Given a norm $\rho(\bar{u})$ in $\mathcal{R}^n$, the optimum scale factor (κ_{opt}) at which the MRE becomes minimum is*

---

[5] $D_t^{(n)} \equiv < 1, 2, 3, \ldots, t, \underbrace{t, \ldots, t}_{n-t} >.$

$$\kappa_{opt} = \frac{2}{\frac{1}{d_{max}} + \frac{1}{b_{min}}} \tag{4.74}$$

where $d_{max} \geq 1$ is the maximum Euclidean distance of a vertex of the unit hypersphere of $\rho(.)$ from the center, and $b_{min} \leq 1$ is the minimum distance of a surface point of the same hypersphere from the center. With κ_{opt} the scale adjusted MRE (MRE_{sc}) is given by the following expression:

$$MRE_{sc} = \frac{d_{max} - b_{min}}{d_{max} + b_{min}} \tag{4.75}$$

Proof If the distance values of ρ are scaled by κ, corresponding d_{max} and b_{min} become $\frac{d_{max}}{\kappa}$, and $\frac{b_{min}}{\kappa}$, respectively. We also apply the constraint of scaling such that $\frac{b_{min}}{\kappa} \leq 1 \leq \frac{d_{max}}{\kappa}$. At optimum scale factor κ_{opt} for minimum MRE, it should satisfy the following:

$$1 - \frac{\kappa_{opt}}{d_{max}} = \frac{\kappa_{opt}}{b_{min}} - 1$$

From the above, we get the expression given in Eq. 4.74. By putting the value of κ_{opt} in Eq. 4.75, we get the expression for scaled adjusted MRE as given in Theorem 4.32. $\qquad\square$

The above theorem is not applicable if $d_{max} < 1$ or $b_{min} > 1$.

Corollary 4.18 *For $CWD_{eu}^{(n)}$ the optimum scale factor is given by*

$$\kappa_{opt} = \frac{2}{1 + \sqrt{\sum_{i=1}^{n} \left(\sqrt{i} - \sqrt{i-1}\right)^2}} \tag{4.76}$$

and the MRE with that scale factor is given by

$$MRE_{sc} = 1 - \frac{2}{1 + \sqrt{\sum_{i=1}^{n} \left(\sqrt{i} - \sqrt{i-1}\right)^2}} = 1 - \kappa_{opt} \tag{4.77}$$

Corollary 4.19 *For $WtD_{isr}^{(n)}$ the optimum scale factor is given by*

Table 4.12 MRE and scale adjusted MRE of distances in different dimensions

Distance (2D)	MRE	MRE_{sc}	Scale	Distance (3D)	MRE	MRE_{sc}	Scale	Distance (4D)	MRE	MRE_{sc}	Scale
$CWD_{eu}^{(2)}$	0.082	**0.040**	0.960	$CWD_{eu}^{(3)}$	0.128	**0.060**	0.940	$CWD_{eu}^{(4)}$	0.160	**0.074**	0.926
$\frac{1}{3} < 3, 4 >$	0.057	0.056	1.002	$\frac{1}{3} < 3, 4, 5 >$	0.106	0.079	0.976	$\frac{1}{3} < 3, 4, 5, 6 >$	0.155	0.101	0.954
$CWD_{opt}^{(2)}$	0.045	0.043	1.002	$CWD_{opt}^{(3)}$	0.074	0.069	1.005	$CWD_{opt}^{(4)}$	0.095	0.089	1.008
$CWD_{opt*}^{(2)}$	0.060	0.052	0.993	$CWD_{opt*}^{(3)}$	0.094	0.082	0.989	$CWD_{opt*}^{(4)}$	0.119	0.106	0.989
$CWD_{umse}^{(2)}$	0.052	**0.040**	1.013	$CWD_{umse}^{(3)}$	0.106	0.071	1.040				
$< 1, 1 >$	0.293	0.172	1.172	$< 1, 1, 1 >$	0.423	0.268	1.268	$< 1, 1, 1, 1 >$	0.500	0.333	1.333
$< 1, 2 >$	0.414	0.172	0.828	$< 1, 2, 3 >$	0.732	0.268	0.732	$< 1, 2, 3, 4 >$	1.000	0.333	0.667
$WtD_{isr}^{(2)}$	0.076	**0.040**	1.040	$WtD_{isr}^{(3)}$	0.114	**0.060**	1.060	$WtD_{isr}^{(4)}$	0.138	**0.074**	1.074

$$\kappa_{opt} = \frac{2}{1 + \frac{1}{\eta}} = \frac{2}{1 + \frac{1}{\sqrt{\sum_{i=1}^{n} \left(\sqrt{i} - \sqrt{i-1}\right)^2}}} \tag{4.78}$$

In the above, η is given by Eq. (4.61). The MRE with that scale factor is given by

$$MRE_{sc} = 1 - \frac{2}{\eta + 1} = 1 - \frac{2}{1 + \sqrt{\sum_{i=1}^{n} \left(\sqrt{i} - \sqrt{i-1}\right)^2}} \tag{4.79}$$

Hence, the scale adjusted MREs of $WtD_{isr}^{(n)}$ and $CWD_{eu}^{(n)}$ are the same.

In Table 4.12, values of MRE, optimally scale adjusted MRE and the optimum scale are shown for a few CWDs in 2-D, 3-D, and 4-D. We have also shown those values for $WtD_{isr}^{(n)}$ norms. We may observe that, as expected, the lowest scale adjusted MREs are achieved by optimally scaled $CWD_{eu}^{(n)}$. This corroborates with the analysis presented in [2]. As we noted in Corollary 4.19, $WtD_{isr}^{(n)}$ achieves also the same performance after being optimally adjusted by a scale.

4.6 Concluding Remarks

Geometric approaches of studying proximity to Euclidean metric provide useful insights into the similarities of hyperspheres of digital distances with Euclidean metrics. We observe that hyperspheres of distance functions which show good proximity in analytical measures have high similarities with the features of those of Euclidean norms. Some of the results, which are obtained analytically, can also be elegantly derived exploiting the properties of hyperspheres. In the next chapter, we discuss more applications of these techniques for analyzing linear combinations of distance functions in approximating Euclidean metrics.

Chapter 5
Linear Combination of Digital Distances

In Chap. 2, it is observed that a linear combination of a number of metrics is also a metric. Hence, linear combinations of digital distances are also considered for approximating the Euclidean distance. In this chapter, we discuss how a good combination could be obtained from the properties of distance functions, which are reviewed in previous chapters.

5.1 Approximation by a Linear Combination of Overestimated and Underestimated Norms

Let us consider first a linear combination of overestimated and underestimated norms (Definitions 4.13 and 4.14). We denote them by ρ_h and ρ_l, respectively. In that case, at every point $\bar{u} \in \mathscr{R}^n$, $\rho_l(\bar{u}) \leq E^{(n)} \leq \rho_h(\bar{u})$. Intuitively a weighted sum of these two norms, such as $a\rho_l(\bar{u}) + b\rho_h(\bar{u})$, $a, b \in \mathscr{P}$, may bring the value closer to the Euclidean norm. Moreover, as ρ_h and ρ_l are metrics, their linear combination is also a metric (Theorem 2.18). Let us denote the linear combination as

$$D_{a,b}(\cdot; \rho_l, \rho_h) \equiv a\rho_l + b\rho_h$$

In this section we discuss how a suitable set of weights (or coefficients of the combination), a and b, could be obtained for improving the approximation of the Euclidean norm by $D_{a,b}$.

© The Author(s), under exclusive license to Springer Nature Singapore Pte Ltd. 2020   103
J. Mukhopadhyay, *Approximation of Euclidean Metric by Digital Distances*,
https://doi.org/10.1007/978-981-15-9901-9_5

5.1.1 Least Squares Estimation (LSE) of a Linear Combination

In [44, 67], the least squares estimation (LSE) method has been used for determining optimal values of a and b by minimizing the mean squared deviation from the Euclidean norm over a finite set of sampled points. In this context, we define the mean squared error (MSE) of a norm with respect to the Euclidean norm as follows:

Definition 5.1 The mean squared error (MSE) of values of a norm ρ from values of the Euclidean norm is defined as

$$\mathscr{E}_{\bar{u}\in\mathscr{R}^n}(|\rho(\bar{u}) - E^{(n)}(\bar{u})|^2)$$

where $\mathscr{E}_S(x)$ denotes expectation of a random variable x over a set S.

By using the method of least squares estimation (LSE), we minimize $MSE(D_{a,b})$ as given in Definition 5.1, and compute the coefficients a and b as follows:

$$a = \frac{\mathscr{E}_{\bar{u}\in\mathscr{R}^n}(E^{(n)}(\bar{u})\rho_l(\bar{u}))\mathscr{E}_{\bar{u}\in\mathscr{R}^n}(\rho_h^2(\bar{u})) - \mathscr{E}_{\bar{u}\in\mathscr{R}^n}(E^{(n)}(\bar{u})\rho_h(\bar{u}))\mathscr{E}_{\bar{u}\in\mathscr{R}^n}(\rho_l(\bar{u})\rho_h(\bar{u}))}{\mathscr{E}_{\bar{u}\in\mathscr{R}^n}(\rho_l^2(\bar{u}))\mathscr{E}_{\bar{u}\in\mathscr{R}^n}(\rho_h^2(\bar{u})) - (\mathscr{E}_{\bar{u}\in\mathscr{R}^n}(\rho_l(\bar{u})\rho_h(\bar{u})))^2}$$

$$(5.1)$$

$$b = \frac{\mathscr{E}_{\bar{u}\in\mathscr{R}^n}(E^{(n)}(\bar{u})\rho_h(\bar{u}))\mathscr{E}_{\bar{u}\in\mathscr{R}^n}(\rho_l^2(\bar{u})) - \mathscr{E}_{\bar{u}\in\mathscr{R}^n}(E^{(n)}(\bar{u})\rho_l(\bar{u}))\mathscr{E}_{\bar{u}\in\mathscr{R}^n}(\rho_l(\bar{u})\rho_h(\bar{u}))}{\mathscr{E}_{\bar{u}\in\mathscr{R}^n}(\rho_l^2(\bar{u}))\mathscr{E}_{\bar{u}\in\mathscr{R}^n}(\rho_h^2(\bar{u})) - (\mathscr{E}_{\bar{u}\in\mathscr{R}^n}(\rho_l(\bar{u})\rho_h(\bar{u})))^2}$$

$$(5.2)$$

5.1.2 Choosing a Suitable Pair of Norms

Given a pair of norms, the LSE method provides a set of optimal weights for their linear combination so that the mean squared deviation from values of the Euclidean norm gets minimized. In this case, from intuition, we consider a combination of overestimated and underestimated norms for approximating the Euclidean norm. In Chap. 4, we consider a list of overestimated norms, which include $d_1^{(n)}$ (or L_1), $CWD_{eu}^{(n)}$ and a set of t-cost norms, $\{D_t^{(n)}|\lceil\sqrt{n}\rceil \le t \le n\}$. Likewise there are a few possible underestimated norms such as $d_n^{(n)}$ (or L_∞), and $WtD_{isr}^{(n)}$. Hence, there are a few candidate pairs, out of which we may choose an appropriate combination providing a good approximation of the Euclidean norm. We represent a candidate pair of norms for the linear combination as (ρ_l, ρ_h), where ρ_l and ρ_h are underestimated and overestimated norms, respectively. To select a candidate pair, it is possible to evaluate them using empirical error measures. In Chap. 3, we discuss a few such measures such as ENAE, EMNAE, EARE, and EMRE in low-dimensional spaces (Sect. 3.7.2). These measures aggregate the error statistics over a finite set of points.

However, as dimension increases, it is difficult to apply those definitions, as this set includes all the points in the all positive hyperoctant of a bounded Euclidean hypersphere centering origin. This increases the size of the set of points exponentially with increasing dimension. Moreover, in those measures error values are tested on points lying along a few discrete directions. It introduces a bias in a measure. For example, in [14], it is observed that EMRE values are optimistic. In view of this, it is proposed to use random sampling of points in the space, and with increasing dimension, the size of the sample should also increase. As the relative error of a norm remains the same in all the points having the same direction from the origin, it is sufficient to consider the distribution of these values on points lying on the surface of an Euclidean hypersphere of unit radius.[1] Using this principle, we modify the definitions of empirical error measures for any arbitrary dimension as follows:

Definition 5.2 Let us denote the set of points lying on the $(n-1)$-dimensional surface of a hypersphere of unit radius of $E^{(n)}$ as $\mathscr{S}_2^{(n-1)}$. Given a finite set of points $S = \{\bar{u} | \bar{u} \in \mathscr{S}_2^{(n-1)}\}$, the *empirical average relative error on sampling* (EARES) and the *empirical maximum relative error on sampling* (EMRES) of a norm ρ are defined as

$$EARES(\rho) = \frac{1}{|S|} \sum_{\bar{u} \in S} |1 - \rho(\bar{u})| \tag{5.3}$$

$$EMRES(\rho) = \max_{\bar{u} \in S} (|1 - \rho(\bar{u})|) \tag{5.4}$$

In addition to above errors, we also use the root mean squared error from the respective values of the Euclidean norm, whose square (MSE) is minimized to provide the linear combination of a pair of norms.

Definition 5.3 With the set of points S as formed by random sampling of points lying on Euclidean hypersphere of unit radius with uniform probability, the *root mean squared error* (RMSE) of the norm from the Euclidean norm is given by the following expression:

$$RMSE(\rho) = \sqrt{\frac{1}{|S|} \sum_{\bar{u} \in S} (1 - \rho(\bar{u}))^2} \tag{5.5}$$

For computing above measures, it is checked whether there is a convergence of error values with increasing size of the sample size. In an experiment [44], it is reported that the values of EARES and RMSE get converged for all the candidate pairs. However for a linear combination of L_∞ ($d_n^{(n)}$) and $D_t^{(n)}$ EMRES values do not converge beyond dimension 4 within a reasonable number of points (between

---

[1] An algorithm for generating points on a surface of a unit Euclidean hypersphere in n-D with uniform probability is discussed in [51].

10^7 to 10^8). Other than empirical measures, for some combinations, we also ontain theoretical MREs. The following theorem provides MRE of linear combinations of L_∞ and $D_t^{(n)}$:

Theorem 5.1 *For the linear combination norm $D_{a,b}^{(n)}(\cdot; L_\infty, D_t^{(n)})$, the MRE is given by the following expression:*

$$MRE(D_{a,b}^{(n)}(\cdot; L_\infty, D_t^{(n)})) = \max\left(\left|1 - \frac{1}{\max\left(\max\limits_{1\leq k<t}\left(\frac{\sqrt{k}}{a+kb}\right), \frac{\sqrt{n}}{a+tb}\right)}\right|, \left|\sqrt{(a+b)^2 + (t-1)b^2} - 1\right|\right)$$

(5.6)

Proof $D_{a,b}^{(n)}(\cdot; L_\infty, D_t^{(n)})$ can also be represented by the CWD $< a + b, a + 2b, \ldots,$ $a + tb, \ldots, a + tb >$. We obtain the coordinates of the vertices of a unit hypersphere of the norm from Theorem 4.12 and Corollary 4.5. Thus, the maximum distance (d_{max}) of a vertex from the origin is given by

$$d_{max} = \max\left(\max_{1\leq k<t}\left(\frac{\sqrt{k}}{a+kb}\right), \frac{\sqrt{n}}{a+tb}\right)$$

(5.7)

Again, using Theorem 4.31, we compute the shortest distance (b_{min}) from the origin to a surface point on its hyperplane as follows:

$$b_{min} = \frac{1}{\sqrt{(a+b)^2 + (t-1)b^2}}$$

(5.8)

We apply Eqs. (5.7) and (5.8), in Theorem 4.32. This gives us the expression for the MRE.   □

Corollary 5.1 *For the linear combination of L_∞ and L_1 [67] when t equals n, the expression is simplified as follows:*

$$MRE(D_{a,b}^{(n)}(\cdot; L_\infty, L_1)) = \max\left(\left|1 - \frac{1}{\max\limits_{1\leq k\leq n}\left(\frac{\sqrt{k}}{a+kb}\right)}\right|, \left|\sqrt{(a+b)^2 + (n-1)b^2} - 1\right|\right)$$

(5.9)

In Table 5.1, different error measures for optimal linear combinations of norms are shown for dimensions ranging from 2 to 8. We have shown error values of three candidate pairs in an n-dimensional real space, namely, (L_∞, L_1), $(L_\infty, D_{\lceil\sqrt{n}\rceil}^{(n)})$, and $(WtD_{isr}^{(n)}, CWD_{eu}^{(n)})$. In 2D, $D_{\lceil\sqrt{2}\rceil}^{(2)}$ is the same L_1 norm. The values of a and b for each combination are also provided in the table. As observed in [44], the combination of L_∞ and D_t does not have any major advantage for its use in place of (L_∞, L_1) [67], which is also computationally less expensive. In Table 5.1, we also find that

Table 5.1 Performance of linear combination of norms

Dimension	Norm-pairs	a	b	RMSE (%)	EARES (%)	EMRES/MRE (%)
2	(L_∞, L_1)	0.55	0.39	2.33	2.00	5.26
	(WtD_{isr}, CWD_{eu})	0.51	0.47	**0.59**	**0.50**	**1.29**
3	(L_∞, L_1)	0.55	0.36	2.90	2.39	9.99
	$(L_\infty, D_2^{(3)})$	0.35	0.55	4.06	3.22	16.49
	(WtD_{isr}, CWD_{eu})	0.50	0.48	**0.67**	**0.56**	**2.12**
4	(L_∞, L_1)	0.54	0.34	3.17	2.57	13.64
	$(L_\infty, D_2^{(4)})$	0.19	0.67	5.33	4.27	23.27
	(WtD_{isr}, CWD_{eu})	0.50	0.47	**0.73**	**0.60**	**2.63**
5	(L_∞, L_1)	0.53	0.32	3.32	2.67	16.60
	$(L_\infty, D_3^{(5)})$	0.32	0.49	3.24	2.49	20.25
	(WtD_{isr}, CWD_{eu})	0.50	0.47	**0.76**	**0.63**	**2.85**
6	(L_∞, L_1)	0.53	0.31	3.40	2.73	19.07
	$(L_\infty, D_3^{(6)})$	0.23	0.54	3.89	2.99	24.44
	(WtD_{isr}, CWD_{eu})	0.51	0.46	**0.74**	**0.64**	**3.09**
7	(L_∞, L_1)	0.53	0.29	3.44	2.76	21.17
	$(L_\infty, D_3^{(7)})$	0.15	0.59	4.42	3.47	27.70
	(WtD_{isr}, CWD_{eu})	0.51	0.45	**0.80**	**0.65**	**3.16**
8	(L_∞, L_1)	0.52	0.28	3.46	2.77	23.00
	$(L_\infty, D_3^{(8)})$	0.09	0.63	4.92	3.89	30.31
	(WtD_{isr}, CWD_{eu})	0.52	0.45	**0.81**	**0.66**	**3.30**

the linear combination of CWD_{eu} and WtD_{isr} has significantly better performance (highlighted in bold fonts) than other combinations. For them both the RMSE and the EARES are contained within 1%, whereas the EMRES values vary from 1.3 to 3.3%. In a subsequent section, we also discuss theoretical expressions of MREs of their combinations. As noted, the empirical values of the MRE are optimistic than the theoretical values.

5.2 Convex Combination

The LSE method as discussed in the previous section, has one limitation. It does not guarantee that the optimum a and b provide a norm, as they may not be always nonnegative. This motivates to look for convex combination of these norms.

Definition 5.4 A convex combination of overestimated (ρ_h) and underestimated (ρ_l) norms is defined as

$$D_\lambda(\cdot; \rho_l, \rho_h) \equiv (1 - \lambda)\rho_l + \lambda\rho_h, \ \lambda \in [0, 1]$$

In convex combination, we have the following advantage:

$$\rho_l(\bar{u}) \leq D_\lambda(\bar{u}; \rho_l, \rho_h) \leq \rho_h(\bar{u})$$

There are a few reported results on convex combination of (L_∞, L_1) [15, 62]. In [15], the value of λ is taken as $\frac{1}{n-\lfloor\frac{n-2}{2}\rfloor}$ for a good approximation of Euclidean norm. In [62], an expression for deriving optimal λ to obtain the minimum MRE for any dimension has been obtained by solving the following quartic (fourth order) equation of λ.

$$1 - 2\sqrt{\lambda - \lambda^2} = \sqrt{1 + \lambda^2(n-1)} - 1 \tag{5.10}$$

At the the optimal value of λ, denoted by λ_{mre}, the MRE is given by

$$MRE(D_{\lambda_{mre}}(\bar{u}; L_\infty, L_1)) = 1 - 2\sqrt{\lambda_{mre} - \lambda_{mre}^2} \tag{5.11}$$

From Eq. (5.6), we obtain theoretical MRE for any arbitrary λ for the convex combination of (L_∞, D_t) by setting $a = (1 - \lambda)$, and $b = \lambda$. In particular, for the convex combination of L_∞ and L_1 the expression is simplified as follows:

$$MRE(D_\lambda^{(n)}(\cdot; L_\infty, L_1)) = \max\left(\left|1 - \frac{1}{\max\limits_{1 \leq k \leq n}\left\{\frac{\sqrt{k}}{(1-\lambda)+k\lambda}\right\}}\right|, \left|\sqrt{1 + (n-1)\lambda^2} - 1\right|\right) \tag{5.12}$$

We observe that Eq. (5.10) [62], can be derived from Eq. (5.12), by assuming k is continuous. Hence, the optimal MRE in Eq. (5.11), is lower than the true value for the integral k.

The other advantage of using a convex combination of these norms over the general linear combination is that, the convex combination retains the common points of both the hyperspheres of radius R, which are lying on the boundary surface of the corresponding Euclidean hypersphere of the same radius. For example, the common points of hyperspheres of L_∞ and L_1 (also $D_t^{(n)}$), which are lying on the Euclidean hypersphere of the same radius (R), are given by $\phi((R, 0, 0, \ldots, 0))$. For this convex combination, the number of such points in n-D is $2n$. We also observe that minimum number of common points between hyperspheres of $CWD_{eu}^{(n)}$ and $WtD_{isr}^{(n)}$ of radius R those lying on the boundary surface of Euclidean sphere is $3^n - 1$. These are also the vertices of the hypersphere of CWD_{eu} (refer to Corollary 4.6). In Fig. 5.1, we have shown an example of circles (convex polygons) of $CWD_{eu}^{(2)}$ and $WtD_{isr}^{(2)}$ meeting at the circumference of an Euclidean circle of the same radius. There are $3^2 - 1 = 8$ such meeting points. We may note that The convex polygon enclosed by the Euclidean circle is the circle of $CWD_{eu}^{(2)}$, and that of $WtD_{isr}^{(2)}$ encloses the Euclidean circle. The circle of the convex combination lies in the interiors of triangular beads around the circumference of the Euclidean circle.

Fig. 5.1 Common
intersecting points of
Euclidean circle (shown in
red color) with circles of
$CWD_{eu}^{(2)}$ (inside the
Euclidean circle shown in
blue color) and $WtD_{isr}^{(2)}$
(outside the Euclidean circle
shown in green color)

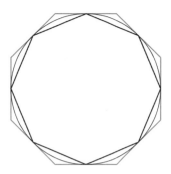

5.2.1 The LSE Method for Finding Optimal λ

We can compute optimal λ by minimizing the MSE of the convex combination. Using
the LSE method we get the optimal solution of λ as follows:

$$\lambda_{lse} = \frac{\mathscr{E}_{\bar{u} \in \mathscr{R}^n}(E^{(n)}(\bar{u})(\rho_h(\bar{u}) - \rho_l(\bar{u}))) - \mathscr{E}_{\bar{u} \in \mathscr{R}^n}(\rho_l(\bar{u})(\rho_h(\bar{u}) - \rho_l(\bar{u})))}{\mathscr{E}_{\bar{u} \in \mathscr{R}^n}(\rho_h(\bar{u})(\rho_h(\bar{u}) - \rho_l(\bar{u}))) - \mathscr{E}_{\bar{u} \in \mathscr{R}^n}(\rho_l(\bar{u})(\rho_h(\bar{u}) - \rho_l(\bar{u})))} \quad (5.13)$$

It is guaranteed that the LSE method provides a solution in $(0, 1)$, thus without
violating the property of convex combination as stated in the following theorem:

Theorem 5.2 *If* $\forall \bar{u} \in \mathscr{R}^n$, $\rho_l(\bar{u}) \leq L_2(\bar{u}) \leq \rho_h(\bar{u})$, *then* $\lambda_{lse} \in (0, 1)$

Proof Let us consider the expression in RHS of Eq. (5.13). As $E^{(n)}(\bar{u}) \geq \rho_l(\bar{u})$
and $\rho_h(\bar{u}) \geq \rho_l(\bar{u})$, the numerator and denominator are positive values. Again, as
$\rho_h(\bar{u}) \geq L_2(\bar{u})$, the denominator is greater or equal to the numerator. Hence, the
theorem. □

In Table 5.2, RMSE, EARES, and EMRES values of optimal convex combination of two candidate pairs of norms are shown. For convex combination too,
$(WtD_{isr}^{(n)}, CWD_{eu}^{(n)})$ has the least errors (highlighted by bold fonts) among its peers.

5.3 Linear Combination of WtD and CWD

In previous sections, empirically we find that a linear combination of $WtD_{isr}^{(n)}$ and
$CWD_{eu}^{(n)}$ provides a good approximation of Euclidean norm. In this section, we
explore the properties of a linear combination of any arbitrary WtD and CWD as a pair
of distance functions. We study the properties of their hyperspheres and subsequently
derive the expressions of theoretical MRE values of their combinations, leading to
the choice of good combinations for approximation.

Table 5.2 Performance of convex combination of norms

Dimension	Norm-pairs	λ_{lse}	RMSE (%)	EARES (%)	EMRES/MRE (%)
2	(L_∞, L_1)	0.30	3.55	3.13	8.04
	(WtD_{isr}, CWD_{eu})	0.36	**0.87**	**0.76**	**1.95**
3	(L_∞, L_1)	0.27	4.46	3.92	11.27
	(WtD_{isr}, CWD_{eu})	0.34	**1.97**	**0.92**	**3.05**
4	(L_∞, L_1)	0.25	4.84	4.18	13.33
	(WtD_{isr}, CWD_{eu})	0.34	**1.15**	**0.96**	**3.73**
5	(L_∞, L_1)	0.24	5.01	4.26	14.90
	(WtD_{isr}, CWD_{eu})	0.33	**1.19**	**0.98**	**4.20**
6	(L_∞, L_1)	0.23	5.08	4.26	16.08
	(WtD_{isr}, CWD_{eu})	0.33	**1.21**	**0.99**	**4.57**
7	(L_∞, L_1)	0.22	5.10	4.23	17.26
	(WtD_{isr}, CWD_{eu})	0.33	**1.22**	**0.99**	**4.86**
8	(L_∞, L_1)	0.21	5.09	4.19	18.34
	(WtD_{isr}, CWD_{eu})	0.32	**1.22**	**0.99**	**5.15**

Definition 5.5 A linear combination of two norms, such that one of them belongs to WtD and the other to CWD is defined below [45].

$$WtCWD^{(n)}(\bar{u}; W, \Delta, a, b) = a \cdot WtD^{(n)}(\bar{u}; W) + b \cdot CWD^{(n)}(\bar{u}; \Delta) \quad (5.14)$$

where $a \in \mathscr{P}$ and $b \in \mathscr{P}$, and at least one of them should be nonzero.

We call the above class of norms $WtCWD$. A typical example of a member of this class is a linear combination of $CWD_{eu}^{(n)}$ and $WtD_{isr}^{(n)}$, which is represented by $WtCWD(\bar{u}; \{1, \frac{1}{\sqrt{2}}, \ldots, \frac{1}{\sqrt{n}}\}, \{1, \sqrt{2}, \ldots, \sqrt{n}\}, a, b)$.

5.3.1   WtCWD as Nonlinear Combination of LWDs

Let us consider a class of distance function in the following form [45]:

Definition 5.6

$$MaxLWD^{(n)}(\bar{u}; \Gamma_1, \Gamma_2, \ldots, \Gamma_k) = \max_{i=1}^{k}\{LWD(\bar{u}; \Gamma_i)\} \quad (5.15)$$

where k is a positive integer. We refer to Γ_i as a member LWD of the distance function. The class of distance function is called *MaxLWD*.

We note that WtD (refer to Definition 2.16) is a member of the class MaxLWD, with $\Gamma_t = \{w_t, w_t, \ldots, w_t, \underbrace{0, \ldots, 0}_{n-t \text{ zeroes}}\}$, $1 \leq t \leq n$. Similarly, a WtCWD distance norm can also be expressed in the form of MaxLWD as stated in the following theorem:

Theorem 5.3 *Let the equivalent LWD of* $CWD^{(n)}(\bar{u}; \Delta)$ *be* $LWD^{(n)}(\bar{u}; \Gamma)$, *such that* $\Gamma = \{\gamma_1, \gamma_2, \ldots, \gamma_n\}$, *and* $W = \{w_1, w_2, \ldots, w_n\}$ *be weight vectors of* $WtD^{(n)}(\bar{u}; W)$. *In that case,* $WtCWD^{(n)}(\bar{u}; W, \Delta, a, b)$ *is equivalent to* $MaxLWD^{(n)}(\bar{u}; \Gamma_1, \Gamma_2, \ldots, \Gamma_n)$, *such that*

$$\Gamma_i = \{aw_i + b\gamma_1, aw_i + b\gamma_2, \ldots, aw_i + b\gamma_i, b\gamma_{i+1}, \ldots, b\gamma_n\} \, for \, 1 \leq i \leq n.$$
(5.16)

While using a combination of WtD and CWD, it is required to check CWD's metricity conditions, otherwise they do not satisfy the conditions of being a norm. It may be noted that a WtD by definition satisfies the conditions for being a norm. One of the necessary conditions for CWD being a norm is that it should have weights in nondecreasing order (refer to Theorem 2.3). On the other hand, the weights of a well-behaved WtD, whose hyperspheres are given by Theorem 4.19, should be in a nonincreasing order. We consider a class of WtCWD, called *reciprocal WtCWD* [45], whose one set of weights can be derived from the other set. A typical example of a reciprocal WtCWD is a combination of $WtD_{isr}^{(n)}$ and $CWD_{eu}^{(n)}$. As this combination has been found to provide a good approximation of Euclidean norm, a few more candidates in this class have also been studied in [45], and are found to have good approximation properties.

5.3.2 Reciprocal WtCWD

Let us define the special class of WtCWD, which satisfies the reciprocity property between the weights of WtD and CWD in the following manner.

Definition 5.7 $WtCWD^{(n)}(\bar{u}; W, \Delta, a, b)$ is called reciprocal WtCWD if for all t, $w_t = \frac{1}{\delta_t}$, where $w_t \in W$, and $\delta_t \in \Delta$. The respective pairing of WtD and CWD is called a reciprocal pair.

It may be noted that reciprocal weights of a well-behaved WtD do not necessarily satisfy norm conditions of a CWD. Hence, a few examples of reciprocal pairs, which also satisfy properties of a norm, are provided in Table 5.3, from [45]. In the above table, $WtCWD_{isr\_eu}^{(n)}$ is a norm as we know $WtD_{isr}^{(n)}$ and $CWD_{eu}^{(n)}$ are norms, and they provide underestimation [42] and overestimation [43], of the Euclidean norm at every point in the space, respectively.

The WtD component of $WtCWD_{sub\_rec}^{(n)}$ is defined in Sect. 2.5.1, which is called *simple upper bound optimized* WtD [42]. It can be easily shown that the corresponding CWD component satisfies also conditions of a metric. For $WtCWD_{rec\_lwdgp}^{(n)}$, the weights of the equivalent LWD of the CWD component decrease in geometric progression, and hence it is also a norm (Theorem 2.2).

Table 5.3 Examples of a few norms formed by reciprocal pairs of WtD and CWD

Name	$w_t \in W$	$\delta_t \in \Delta$
$WtCWD^{(n)}_{isr\_eu}$	$\frac{1}{\sqrt{t}}$	$\sqrt{t}$
$WtCWD^{(n)}_{sub\_rec}$	$w_t = \begin{cases} \frac{2}{\sqrt{t}+\frac{t}{\sqrt{n}}} & t \leq \sqrt{n}. \\ \frac{2}{1+\sqrt{t}} & \sqrt{n} \leq t \leq n. \end{cases}$	$\delta_t = \begin{cases} \frac{\sqrt{t}+\frac{t}{\sqrt{n}}}{2} & t \leq \sqrt{n}. \\ \frac{1+\sqrt{t}}{2} & \sqrt{n} \leq t \leq n. \end{cases}$
$WtCWD^{(n)}_{rec\_lwdgp}$	$\frac{2^{t-1}}{2^t-1}$	$\frac{2^{t-1}}{2^t-1}$

5.4 Vertices of Hyperspheres of WtCWD

In this section we discuss how vertices of a hypersphere of a WtCWD norm could be computed. The principle behind this approach is to express a WtCWD in the form of an equivalent MaxLWD norm. In that case, the hypersphere of the WtCWD can also be formed by the intersection of hyperspheres of the member LWDs of the corresponding MaxLWD function.

Theorem 4.12 computes vertices of a hypersphere of an LWD with the set of weights Γ [43]. By computing the intersection of hypespheres of member LWDs, we get the vertices of hypersphere of the respective WtCWD norm. Following theorem states the result of this computation [45]:

Theorem 5.4 *Let* $W = \{w_i | 1 \leq i \leq n\}$, $\Delta = \{\delta_i | 1 \leq i \leq n\}$, *and* $\Gamma = \{\gamma_i | 1 \leq i \leq n\}$ *such that* $\gamma_1 = \delta_1$, *and* $\gamma_i = \delta_i - \delta_{i-1}$, *for* $1 < i \leq n$. *Let us also assume that* W *follows the weight assignment of a well-behaved WtD, i.e.,* $w_1 \geq w_2 \geq \ldots \geq w_n > 0$. *For* $WtCWD^{(n)}(\bar{u}; W, \Delta, a, b)$, *the vertices of a hypersphere of radius* R *are given by the following sets:*

1. $\left\{ U_t | U_t = \phi \left(\frac{R}{f_t}, \frac{R}{f_t}(\frac{w_1}{w_2} - 1), \frac{R}{f_t}(\frac{w_1}{w_3} - \frac{w_1}{w_2}), \ldots, \frac{R}{f_t}(\frac{w_1}{w_t} - \frac{w_1}{w_{t-1}}), 0, \ldots, 0 \right), 2 \leq t \leq n \right\}$,
   where

$$f_t = (aw_1 + b\gamma_1) + b\sum_{i=2}^{t} \gamma_i(\frac{w_1}{w_i} - \frac{w_1}{w_{i-1}}) \ for \ 1 < t \leq n \qquad (5.17)$$

These vertices are called Type-I *vertices.*

2. $\left\{ V_t \ \middle| \ V_t = \phi \left(\underbrace{\frac{R}{g_t}, \ldots, \frac{R}{g_t}}_{t}, \underbrace{0, \ldots, 0}_{n-t} \right) , \ for \ 1 \leq t \leq n, \right\}$
   where

$$g_t = \max \{aw_1 + b\delta_t, 2aw_2 + b\delta_t, \ldots, taw_t + b\delta_t\}, \ for \ 1 \leq t \leq n \qquad (5.18)$$

These vertices are called Type-II *vertices.*

Proof Let us consider hyperspheres of LWDs in the MaxLWD equivalent form on the distance function.

Case I: The vertices are formed by intersection of any t hyperspheres, for $2 \leq t \leq n$. It implies that a vertex $(\bar{u})$ is a solution of any t equations as given below

$$LWD^{(n)}\bar{u}; \Gamma_{i_1}) = R$$
$$LWD^{(n)}(\bar{u}; \Gamma_{i_2}) = R$$
$$\vdots \qquad\qquad\qquad (5.19)$$
$$LWD^{(n)}(\bar{u}; \Gamma_{i_t}) = R$$

for some $i_1 < i_2 < \cdots < i_t$. At the same time, the following constraints should be satisfied by the solution:

$$LWD^{(n)}(\bar{u}; \Gamma_k) < R, \forall k \in \{1, 2, \ldots, n\} - \{i_1, i_2, \ldots, i_t\}. \qquad (5.20)$$

Without loss of generality, let us put the constraint on $\bar{u}$ as follows: $u_1 \geq u_2 \geq u_3 \geq \cdots \geq u_n \geq 0$.

As there are t equations, at a vertex a point should have a coordinate value as high as possible in a dimension. In view of that, we consider those solutions for which last $(n - t)$ coordinate values are set to 0. Hence, the form of the solution of the vertex is further restricted as follows: $u_1 \geq u_2 \geq u_3 \geq \cdots \geq u_t > u_{t+1} = u_{t+2} = \cdots = u_n = 0$.

From the above form, we observe that t equations in Eq. (5.19), have t unknowns, and hence it has a single solution. However, not every system has a solution that satisfies the constraint of coordinate ordering and values as mentioned above. It is shown that for each $2 \leq t \leq n$, there is only one set of t equations, which has a solution in the desired form (refer to the lemmas in Appendix of [45]). This unique set of equations is given below

$$\begin{aligned}(aw_1 + b\gamma_1)u_1 + b\gamma_2 u_2 + b\gamma_3 u_3 + \cdots + b\gamma_t u_t &= R \\ (aw_2 + b\gamma_1)u_1 + (aw_2 + b\gamma_2)u_2 + b\gamma_3 u_3 + \cdots + b\gamma_t u_t &= R\end{aligned}$$
$$\vdots$$
$$(aw_t + b\gamma_1)u_1 + (aw_t + b\gamma_2)u_2 + (aw_t + b\gamma_3)u_3 + \cdots + (aw_t + b\gamma_t)u_t = R$$
$$(5.21)$$

By subtracting top row of equation from the rest, we transform this set into the following form:

$$\begin{bmatrix} (aw_1 + b\gamma_1) & b\gamma_2 & b\gamma_3 & \ldots & b\gamma_t \\ (aw_2 - aw_1) & aw_2 & 0 & \ldots & 0 \\ \vdots & & & & \\ (aw_t - aw_1) & aw_t & aw_t & \ldots & aw_t \end{bmatrix} \begin{bmatrix} u_1 \\ u_2 \\ \vdots \\ u_t \end{bmatrix} = \begin{bmatrix} R \\ 0 \\ \vdots \\ 0 \end{bmatrix}$$

By solving the above the corresponding vertex is obtained as $(\frac{R}{f_t}, \frac{R}{f_t}(\frac{w_1}{w_2} - 1), \frac{R}{f_t}(\frac{w_1}{w_3} - \frac{w_1}{w_2}), \frac{R}{f_t}(\frac{w_1}{w_4} - \frac{w_1}{w_3}), \ldots, \frac{R}{f_t}(\frac{w_1}{w_t} - \frac{w_1}{w_{t-1}}), 0, 0, \ldots, 0)$, where

$$f_t = (aw_1 + b\gamma_1) + b\sum_{i=2}^{t} \gamma_i \left(\frac{w_1}{w_i} - \frac{w_1}{w_{i-1}}\right).$$

Case II: Other set of vertices is those of an individual member CWD, which lie in the innermost core of the intersection of all the member CWDs. From Corollary 4.5, vertices of ith member CWD is given as

$$\phi(\tfrac{R}{\delta_{i,1}}, 0, \ldots, 0), \ldots, \phi(\tfrac{R}{\delta_{i,t}}, \tfrac{R}{\delta_{i,t}}, \ldots, \tfrac{R}{\delta_{i,t}}, 0, \ldots, 0), \ldots \phi(\tfrac{R}{\delta_{i,n}}, \tfrac{R}{\delta_{i,n}}, \ldots, \tfrac{R}{\delta_{i,n}}),$$

where

$$\delta_{i,t} = \sum_{s=1}^{t} \gamma_{i,s}, \text{ where } \gamma_{i,s} \text{ is the } s\text{th member of } \Gamma_i$$

$$= \min\{i, t\} \cdot a \cdot w_i + b \cdot \delta_t, \text{ for } 1 \le t \le n.$$

As among the member LWDs, these vertices should have minimum values at coordinate locations, we obtain the vertices as shown in the theorem.   □

A few examples of circles and spheres of WtCWD distances in 2-D and 3-D, are shown in Fig. 5.2, and Fig. 5.3, respectively.

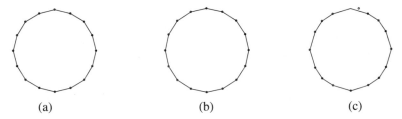

(a)   (b)   (c)

Fig. 5.2 Circles of WtCWD with $a = 0.4, b = 0.6$: **a** $WtCWD^{(2)}_{isr\_eu}$ ($W = \{1, \tfrac{1}{\sqrt{2}}\}, \Delta = \{1, \sqrt{2}\}$), **b** $WtCWD^{(2)}_{sub\_rec}$ ($W = \{1.1716, 0.8284\}, \Delta = \{0.8536, 1.2071\}$), and **c** $WtCWD^{(2)}_{rec\_lwdgp}$ ($W = \{1, \tfrac{2}{3}\}, \Delta = \{1, \tfrac{3}{2}\}$)

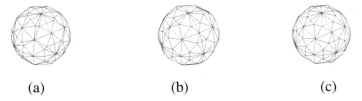

(a)   (b)   (c)

Fig. 5.3 Spheres of WtCWD with $a = 0.4, b = 0.6$:: **a** $WtCWD^{(3)}_{isr\_eu}$ ($W = \{1, \tfrac{1}{\sqrt{2}}, \tfrac{1}{\sqrt{3}}\}, \Delta = \{1, \sqrt{2}, \sqrt{2}\}$), **b** $WtCWD^{(3)}_{sub\_rec}$ ($W = \{1.2679, 0.8284, 0.7321\}, \Delta = \{0.7887, 1.2071, 1.3660\}$), and **c** $WtCWD^{(2)}_{rec\_lwdgp}$ ($W = \{1, \tfrac{2}{3}, \tfrac{4}{7}\}, \Delta = \{1, \tfrac{3}{2}, \tfrac{7}{4}\}$)

5.4.1 MRE of WtCWD

From Theorem 5.4, we can compute vertices of a hypersphere of WtCWD. This is useful in computing MRE of the norm by applying Theorem 4.32. This is stated in the following theorem:

Theorem 5.5 *The MRE of the norm $WtCWD^{(n)}(\bar{u}; W, \Delta, a, b)$, which follows the same representation as described in Theorem 5.4 is given by the following:*

$$MRE(WtCWD^{(n)}(\bar{u}; W, \Delta, a, b)) = \max\left(\left|1 - \frac{1}{d_{max}}\right|, \left|1 - \frac{1}{b_{min}}\right|\right) \quad (5.22)$$

where

$$d_{max} = \sqrt{\max\left\{\max_{t=2}^{n}\left\{\frac{1}{f_t^2}\left(1 + \sum_{i=2}^{t}\left(\frac{w_1}{w_i} - \frac{w_1}{w_{i-1}}\right)^2\right)\right\}, \max_{t=1}^{n}\left\{\frac{t}{g_t^2}\right\}\right\}}, \quad (5.23)$$

$$b_{min} = \min_{t=1}^{n}\left\{\frac{1}{\sqrt{\sum_{i=1}^{t}(aw_t + b\gamma_i)^2 + \sum_{i=t+1}^{n}(b\gamma_i)^2}}\right\}, \quad (5.24)$$

$$f_t = (aw_1 + b\gamma_1) + b\sum_{i=2}^{t}\gamma_i\left(\frac{w_1}{w_i} - \frac{w_1}{w_{i-1}}\right) \text{ for } 1 < t \le n, \quad (5.25)$$

and

$$g_t = \max\left\{aw_1 + b\sum_{i=1}^{t}\gamma_i, 2aw_2 + b\sum_{i=1}^{t}\gamma_i, 3aw_3 + b\sum_{i=1}^{t}\gamma_i, \ldots, taw_t + b\sum_{i=1}^{t}\gamma_i\right\}. \quad (5.26)$$

$$\text{for } 1 \le t \le n.$$

Proof From Theorem 5.4, we obtain d_{max} (as defined in Theorem 4.32) of WtCWD as follows:

$$d_{max} = \sqrt{\max\left\{\max_{t=2}^{n}\left\{\frac{1}{f_t^2}\left(1 + w_1^2\sum_{i=2}^{t}\left(\frac{1}{w_i} - \frac{1}{w_{i-1}}\right)^2\right)\right\}, \max_{t=1}^{n}\left\{\frac{t}{g_t^2}\right\}\right\}} \quad (5.27)$$

Let us consider a hypersphere of a WtCWD centering at the origin. As it is a convex polytope, the shortest distance of a point in the surface of the hypersphere is the minimum perpendicular distance (the distance along the normal of a hyperplane) of its hyperplanes from the origin. This can also be computed by computing this shortest distance for each member LWDs and by taking the minimum of them. From Theorem 4.31, we compute this shortest distance from the origin to a boundary hyperplane of a hypersphere of radius R, and by considering all the member LWDs we obtain b_{min} (as defined in Theorem 4.32) as follows[2]:

---

[2]It could be equivalently written as:

$$b_{min} = \min_{t=1}^{n} \left\{ \frac{1}{\sqrt{\sum_{i=1}^{t}(aw_t + b\gamma_i)^2 + \sum_{i=t+1}^{n}(b\gamma_i)^2}} \right\} \tag{5.28}$$

$\square$

Corollary 5.2 *From d_{max} and b_{min} of Eq. (5.23) and Eq. (5.24), respectively, we obtain optimum scale and scale adjusted MRE from Theorem 4.35. These are given as follows:*

$$\kappa_{opt} = \frac{2}{\frac{1}{d_{max}} + \frac{1}{b_{min}}} \tag{5.29}$$

$$MRE_{sc} = \frac{d_{max} - b_{min}}{d_{max} + b_{min}} \tag{5.30}$$

The above equations are valid, if $d_{max} \geq 1$ and $b_{min} \leq 1$.

5.5 Optimum Linear Combination by Minimizing MRE

As we are able to compute theoretical value of MRE for a linear combination of WtD and CWD following Theorem 5.5, we may consider finding optimum coefficients a and b in its expression by minimizing the MRE. In Fig. 5.4, we show the functional distributions of MRE values over $a - b$ parameter space for $WtCWD_{isr\_eu}$, $WtCWD_{sub\_rec}$, and $WtCWD_{rec\_lwdgp}$ distances in 2-D and 3-D. We have computed MRE values within the domain $a \times b = [0, 1]^2$. All the values are scaled by 100 in the plots. In all these plots, we observe that there is a saddle ridge in the functional surface, showing a path of local optima. It is interesting to note that this path is almost a straight line and follows the off-diagonal ($a + b = 1$) of the parameter space. In particular, for the distance $WtCWD_{isr\_eu}$ these lines are very close to the off-diagonal both in 2-D and 3-D. This has also been observed in higher dimensions. This provides an empirical justification of our choice of convex combinations for underestimated and overestimated norms for approximating an Euclidean norm. In Fig. 5.5, plots of MRE against the coefficient of convex combination $\lambda(=b)$ are shown for all those distances in different dimensions. We observe a global minimum point in each of these curves, and from there we obtain the optimum MRE and optimum λ of the combination. For observing the trends in higher dimensions, in the figures, plots of MREs are shown in dimensions 10, 100, and 1000 in addition to low dimensional spaces such as in 2-D, 3-D, and 4-D. In Table 5.4, we also show the

---

$$b_{min} = \min_{t=1}^{n} \left\{ \frac{1}{E^{(n)}(\Gamma_t)} \right\}.$$

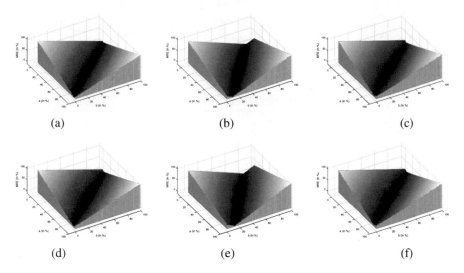

Fig. 5.4 Distribution of MRE in the parameter space $(a - b)$ of linear combination: **a** $WtCWD^{(2)}_{isr\_eu}$ **b** $WtCWD^{(2)}_{sub\_rec}$, **c** $WtCWD^{(2)}_{rec\_lwdgp}$, **d** $WtCWD^{(3)}_{isr\_eu}$ **e** $WtCWD^{(3)}_{sub\_rec}$, and **f** $WtCWD^{(3)}_{rec\_lwdgp}$

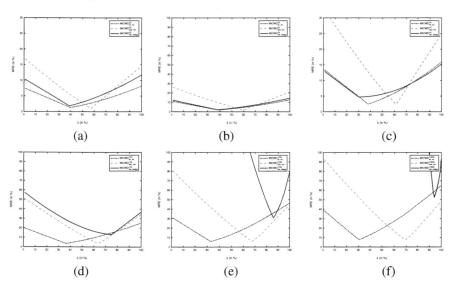

Fig. 5.5 Plots of MRE against λ of convex combinations in different dimensions: **a** 2D, **b** 3D, **c** 4D, **d** 10D **e** 100D, and **f** 1000D

Table 5.4 Optimum λ and optimum MRE values of convex combinations of $WtCWD$s

Distance	2-D		3-D		4-D		10-D		100-D		1000-D	
	λ	MRE (in %)	λ	MRE (in %)	λ	MRE (in %)	λ	MRE (in %)	λ	MRE (in %)	λ	MRE (in %)
$WtCWD_{isr\_eu}$	0.4	1.363	0.39	2.052	0.38	2.468	0.36	**3.821**	0.33	**6.072**	0.31	7.969
$WtCWD_{sub\_rec}$	0.57	**1.060**	0.60	**1.962**	0.61	**2.441**	0.63	3.916	0.67	6.093	0.69	**7.673**
$WtCWD_{rec\_lwdgp}$	0.39	1.884	0.38	2.347	0.30	4.740	0.75	13.206	0.86	31.032	0.94	52.502

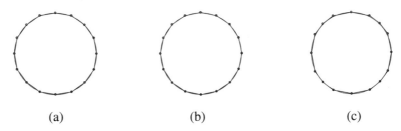

(a)   (b)   (c)

Fig. 5.6 Circles of optimum convex combinations and Euclidean circle of same radius: **a** $WtCWD_{isr\_eu}$, with $\lambda = 0.4$, **b** $WtCWD_{sub\_rec}$ with $\lambda = 0.57$, and **c** $WtCWD_{rec\_lwdgp}$ with $\lambda = 0.39$. Euclidean circles of same radii and centers are also shown in red color with the $WtCWD$ circles

values of optimum MRE and optimum λ from these plots. There is a further possibility of improvement with scale adjustment. However, at these optimum points these improvements, if any, are found to be very marginal providing a scale very close to 1. We observe that all the candidate distances have very low MRE in the optimum convex combination in lower dimensional spaces. Both $WtCWD_{isr\_eu}$ and $WtCWD_{sub\_rec}$ have very low MREs in higher dimensions also, so that even at a very high dimension such as 1000, the optimum MRE values are around 8% only. In low dimensional spaces, $WtCWD_{sub\_rec}$ is found to have the best performances out of these distances. Circles of these distances in 2-D with optimum λ values are shown in Fig. 5.6. They quite closely follow the circumference of the Euclidean circle of the same radius and center, which is shown in red color in the same figure.

5.6   MRE of Sub-classes of WtCWD

WtCWD generalizes some of the classes of distance functions such as CWD, which includes t-cost distances, and WtD including its approximation of HODs of sorted neighborhood sequences and m-neighbor distances. Hence, from Theorem 5.5, we derive MREs of distance functions of all these sub-classes. For example, by setting $a = 0$ and $b = 1$ in Theorem 5.5, we get MRE of any arbitrary CWD norm, which would be the same as stated in Theorem 4.33, in the previous chapter. In the same way, we get the expressions for a well-behaved WtD, which we have not yet discussed in previous chapters. In this section, let us review results related to WtD and its sub-classes as special cases of Theorem 5.5.

5.6.1 MRE for WtD

By setting $a = 1$ and $b = 0$ in Theorem 5.5, we get the expression of MRE for a well-behaved WtD. This is stated in the following theorem:

Theorem 5.6 *MRE of* $Wt D^{(n)}(\bar{u}; W)$ *is given by*

$$MRE(WtD^{(n)}(\bar{u}; W)) = \max\left(\left|1 - \frac{1}{d_{max}}\right|, \left|1 - \frac{1}{b_{min}}\right|\right) \tag{5.31}$$

where

$$d_{max} = \sqrt{\max\left\{\frac{1}{w_1^2} + \sum_{i=2}^{n}\left(\frac{1}{w_i} - \frac{1}{w_{i-1}}\right)^2, \max_{t=1}^{n}\left\{\frac{1}{t.w_t^2}\right\}\right\}}, \tag{5.32}$$

and

$$b_{min} = \sqrt{\min_{t=1}^{n}\left\{\frac{1}{t.w_t^2}\right\}}. \tag{5.33}$$

5.6.2 MRE of Hyperoctagonal Distances Defined by Sorted Neighborhood Sequences

Using the approximation of HODs defined by a sorted neighborhood sequence B by a WtD, we obtain its MRE from Theorem 3.9 [46].

Theorem 5.7 *Consider a sorted neighborhood sequence B, with a fixed vector representation $\Omega = [\omega_1, \omega_2, \ldots, \omega_n]$. Let α_i be the fraction of neighborhood types greater or equal to the type i in the sequence B of length p, so that $\alpha_i = \dfrac{\sum_{t=i}^{n}\omega_t}{p}$. MRE of $\hat{d}_B^{(n)}(\bar{u}; \Omega)$ is given by*

$$MRE(\hat{d}_B^{(n)}(\bar{u}; \Omega)) = \max\left(\left|1 - \frac{1}{d_{max}}\right|, \left|1 - \frac{1}{b_{min}}\right|\right) \tag{5.34}$$

The corresponding expressions for d_{max} and b_{min} are given by

$$d_{max} = \max\left\{\sqrt{1 + \sum_{i=2}^{n}\alpha_i^2}, \max_{t=1}^{n}\left\{\frac{\left(1 + \sum_{i=2}^{t}\alpha_i\right)}{\sqrt{t}}\right\}\right\} \tag{5.35}$$

and

$$b_{min} = \min_{t=1}^{n}\left\{\frac{\left(1 + \sum_{i=2}^{t}\alpha_i\right)}{\sqrt{t}}\right\}. \tag{5.36}$$

Proof In the WtD form of representation, the weights W representing the $\hat{d}_B^{(n)}(\bar{u}; \Omega)$ are given by $w_t = \frac{p}{f_t(p)}$, for $t = 1, 2, \ldots, n$ (refer to Eqs. 4.50, and 4.51). From Eq. 4.51, we get the following:

$$\frac{1}{w_t} = 1 + \sum_{i=2}^{t} \alpha_i \tag{5.37}$$

Hence,

$$\frac{1}{w_t} - \frac{1}{w_{t-1}} = \alpha_t, \text{ for } 2 \le t \le n \tag{5.38}$$

Similarly, $\frac{1}{t.w_t^2}$ reduces to $\dfrac{\left(1 + \sum\limits_{i=2}^{t} \alpha_i\right)^2}{t}$. From the above, we get the expressions in Eqs. (5.34)–(5.36). □

5.6.2.1 For Distances in 2-D

In Das [20], following an analytical approach, an expression of MRE for simple octagonal distances in 2-D is derived. This is discussed in Theorem 3.19. From Theorem 5.7, we get an equivalent expression [46] of Theorem 3.19[3] as given below

$$MRE(\hat{d}_B^{(n)}(\bar{u}; \Omega = \{\omega_1, \omega_2\})) = \max\left\{\left|1 - \frac{1}{\sqrt{1 + \alpha_2^2}}\right|, \left|1 - \frac{\sqrt{2}}{1 + \alpha_2}\right|\right\}, \tag{5.39}$$

where

$$\alpha_2 = \frac{\omega_2}{\omega_1 + \omega_2}$$

The minimum of the MRE values occurs at α_2 satisfying the following equation:

$$1 - \frac{1}{\sqrt{1 + \alpha_2^2}} = \frac{\sqrt{2}}{1 + \alpha_2} - 1$$

The solution of the above equation is at $\alpha_2 = 0.34201$. It corresponds to a neighborhood sequence in 2D such that the ratio of $type$-1 and $type$-2 neighbors is 1.923887. The value of MRE with such a neighborhood sequence is 0.053809. The same value is also obtained in [20]. In Chaps. 3 and 4, we observe that $B = \{1, 1, 2\}$ is a good octagonal distance for approximating Euclidean metric in 2-D. We find that this neighborhood sequence has the ratio of neighborhood types 2:1, which is very close to the optimum point.

5.6.2.2 For Distances in 3-D

In 3-D, MRE is a function of two variables as given below [46].

---

[3]We should note that m in Theorem 3.19 is the same as $1 + \alpha_2$.

$$MRE(\hat{d_B}^{(3)}(\bar{u}; \Omega = \{\omega_1, \omega_2, \omega_3\})) =$$

$$\max \left\{ \left| 1 - \frac{1}{\sqrt{1+\alpha_2^2+\alpha_3^2}} \right|, \left| 1 - \frac{\sqrt{2}}{1+\alpha_2} \right|, \left| 1 - \frac{\sqrt{3}}{1+\alpha_2+\alpha_3} \right| \right\}, \qquad (5.40)$$

where

$$\alpha_2 = \frac{\omega_2 + \omega_3}{\omega_1 + \omega_2 + \omega_3}, \text{ and } \alpha_3 = \frac{\omega_3}{\omega_1 + \omega_2 + \omega_3}.$$

For obtaining optimum MRE, we compute a global minimum in the 2-D parametric space of $\alpha_2 \times \alpha_3$ such that $0 \le \alpha_3 \le \alpha_2 \le 1$. However, as it is computationally intractable, in [46], a gradient descent method from an initial point in the solution space has been applied to arrive at a local minimum. With several trials and errors, a few good candidates in the solution space are identified. A typical such candidate solution is at $\alpha_2 = 0.396356$, and $\alpha_3 = 0.198483$. At that point the MRE is 0.086035, and the proportional ratios among $type$-1, $type$-2, and $type$-3 neighbors are 3.041295 : 0.996930 : 1. A close realization of these parametric values is the HOD with a neighborhood sequence is $\{1, 1, 1, 2, 3\}$. It has an MRE of 0.087129. This finding is corroborated with a previous work [32], where with an independent empirical analysis, this HOD was reported. There exists also a lower MRE value of 0.081475 at $\alpha_2 = 0.307671$, and $\alpha_3 = 0.299993$. In this case, the proportional ratios among different types are computed as 2.307818 : .025592 : 1. The HOD defined by the neighborhood sequence $B = \{1, 1, 3\}$ is a close realization of these values. This distance has an MRE value of 0.095466. $B = \{1, 1, 3\}$ has also low analytical and geometric errors as discussed in Chaps. 3 and 4.

5.6.2.3 For Distances in n-D

It is hard to search a good candidate solution by finding local minima in the multidimensional parametric space. Instead in [46], an enumeration strategy has been adopted to search candidates among the HODs with sorted neighborhood sequences within a fixed length of sequence. From [46], scale adjusted MREs for an optimum neighborhood sequence, within the sequence length L ($2 \le L \le 5$) are shown in Table 5.5. From these tables, some of the interesting empirical observations are noted here [46].

- With increasing L, values of scale adjusted MRE decrease or remain constant.
- A neighborhood sequence of every optimum distance (for $p \le L$) has at least one type-1 neighborhood, whose length is usually the largest among others. The only exception is noted with $L = 2$.
- From other higher order neighborhood types, only a few co-occurs with the type-1. With increasing dimension and increasing L, the order of co-occurred neighborhood type gets higher.

Table 5.5 Optimum HODs with sorted neighborhood sequences with the sequence length p, for $p \leq 2$, $p \leq 3$, $p \leq 4$, and $p \leq 5$, providing minimum MRE

n	$p \leq 2$			$p \leq 3$			$p \leq 4$			$p \leq 5$		
	B	k_{opt}	MRE_{sc}	B	k_{opt}	MRE_{sc}	B	k_{opt}	MRE_{sc}	B	k_{opt}	MRE_{sc}
2	$[1^1 2^1]$	1.056	0.056	$[1^1 2^1]$	1.056	0.056	$[1^1 2^1]$	1.056	0.056	$[1^3 2^2]$	1.032	**0.042**
3	$[1^1 3^1]$	1.101	0.101	$[1^2 3^1]$	1.018	0.079	$[1^2 2^1 3^1]$	1.068	0.068	$[1^2 2^1 3^1]$	1.068	0.068
4	$[1^1 3^1]$	1.101	0.101	$[1^1 3^1]$	1.101	0.101	$[1^2 2^1 4^1]$	1.079	0.079	$[1^2 2^1 4^1]$	1.079	0.079
5	$[1^1 4^1]$	1.139	0.139	$[1^2 5^1]$	1.057	0.121	$[1^2 2^1 5^1]$	1.090	0.090	$[1^2 2^1 5^1]$	1.090	0.090
6	$[1^1 4^1]$	1.139	0.139	$[1^2 5^1]$	1.057	0.121	$[1^2 2^1 6^1]$	1.101	0.101	$[1^2 2^1 6^1]$	1.101	0.101
7	$[1^1 4^1]$	1.102	0.167	$[1^2 6^1]$	1.074	0.139	$[1^2 2^1 7^1]$	1.111	0.111	$[1^3 3^1 7^1]$	1.089	0.108
8	$[1^1 5^1]$	1.172	0.172	$[1^2 6^1]$	1.074	0.139	$[1^2 2^1 8^1]$	1.121	0.121	$[1^3 3^1 8^1]$	1.092	0.114
9	$[1^1 5^1]$	1.172	0.172	$[1^2 7^1]$	1.090	0.156	$[1^2 2^1 8^1]$	1.121	0.121	$[1^2 2^1 8^1]$	1.121	0.121
10	$[1^1 5^1]$	1.136	0.197	$[1^2 7^1]$	1.090	0.156	$[1^2 2^1 9^1]$	1.130	0.130	$[1^3 3^1 10^1]$	1.104	0.127

5.6.3 *m*-Neighbor Distances

The m-neighbor distances are special cases of HODs with a sorted neighborhood sequence of length 1. Hence, they are also represented by WtDs with the following weight assignments [42], as discussed in previous chapters:

$$w_t = \begin{cases} \frac{1}{t}, & \text{for } t < m \\ \frac{1}{m}, & \text{for } t \geq m \end{cases} \tag{5.41}$$

Using the above weight assignments, from Theorem 3.9, the MRE of an m-neighbor distance can be obtained [46].

Theorem 5.8 *The MRE of $d_m^{(n)}$ is given by*

$$MRE(d_m^{(n)}) = \begin{cases} \max\left\{ \left|1 - \frac{1}{\sqrt{m}}\right|, \left|1 - \frac{\sqrt{n}}{m}\right| \right\} & \text{for } m < \sqrt{n}, \\ \left|1 - \frac{1}{\sqrt{m}}\right| & \text{for } m \geq \sqrt{n}. \end{cases} \tag{5.42}$$

Proof By assigning the weights from Eq. 5.41, in Theorem 5.7, we get the following:

$$d_{max} = \sqrt{\max\left\{ 1 + \sum_{2}^{m} 1, \max_{t=m}^{n}\{\frac{m^2}{t}\} \right\}} = \sqrt{m} \tag{5.43}$$

and,

$$b_{min} = \sqrt{\min\left\{ \min_{t=1}^{m}\{t\}, \min_{t=m}^{n}\{\frac{m^2}{t}\} \right\}}$$

The above reduces to

$$b_{min} = \begin{cases} \frac{m}{\sqrt{n}}, & m < \sqrt{n} \\ 1, & m \geq \sqrt{n} \end{cases} \tag{5.44}$$

Using Eqs. 5.43 and 5.44, we get the given expression of MRE. $\qquad\square$

5.6.3.1   Optimum *m*

From Theorem 5.8, the optimum value of m providing minimum MRE in n-D can
also be obtained [46]. For this we need to solve the following equation, assuming m
is continuously varying between 1 and n.

$$\frac{\sqrt{n}}{m} - 1 = 1 - \frac{1}{\sqrt{m}} \qquad (5.45)$$

In Eq. (5.45), the expression in LHS is monotonically decreasing with m till it
becomes 0 at $m = \sqrt{n}$. The expression in the RHS is monotonically increasing from
0 till it reaches $1 - \frac{1}{\sqrt{n}}$ at $m = n$. By solving the above equation, optimum m, m_{opt}
is given by

$$m_{opt} = \frac{1}{16}\left(2 + 8\sqrt{n} + 2\sqrt{1 + 8\sqrt{n}}\right) \qquad (5.46)$$

As m is an integer, the optimum MRE occurs either at $\lfloor m_{opt}\rfloor$ with the MRE value
($\frac{\sqrt{n}}{\lfloor m_{opt}\rfloor} - 1$), or at $\lceil m_{opt}\rceil$ with the value $(1 - \frac{1}{\sqrt{\lceil m_{opt}\rceil}})$, whichever is minimum, the
plots of m_{opt} and MRE for those distance functions are shown in Fig. 5.7, for varying
dimensions (≤ 40). We may note that the plots are shown with continuous variation.
This is to show the trend of the variation. In [27], using a rigorous analysis with the
help of a conjecture a list of optimum m in different dimensions has been provided.
The optimum values of m obtained from Eq. 5.46, mostly corroborate with the results
reported there. There are differences for some dimensions such as in dimensions 7,
8, 18–20, and 33–38, where m_{opt} is found to be one less those reported in that work.
As these values are directly computed from MRE expressions (Eqs. 5.45 and 5.46),
these are accurate. We also note the reduction of MREs due to scale adjustment and
the optimum behavior by scaling the distance values uniformly (Fig. 5.8). Typical
values of MREs, m_{opt} under both scenarios are shown for a few low-dimensional
spaces $(2 - 5)$ in Table 5.6.

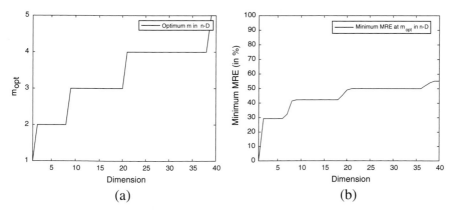

Fig. 5.7 a Optimum m, and **b** MRE for m-neighbor distances at optimum m with varying dimensions

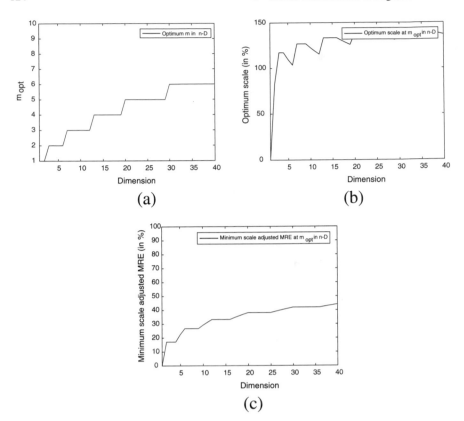

Fig. 5.8 Scale adjusted: **a** Optimum m, **b** optimum scale, and **c** MRE for m-neighbor distances at optimum m with varying dimensions

Table 5.6 Optimum m and MRE values of $d_m^{(n)}$

Dimension	Without scale adjustment		With scale adjustment		
	m_{opt}	MRE	m_{opt}	Scale (κ_{opt})	MRE_{sc}
2	2	0.2929	1	0.8284	0.1716
3	2	0.2929	2	1.1716	0.1716
4	2	0.2929	2	1.1716	0.1716
5	2	0.2929	2	1.0958	0.2251

5.7 Concluding Remarks

Linear combinations of a pair of norms, in particular, convex combinations with an underestimated and an overestimated norms are found to be very effective in approximating an Euclidean norm. In this regard, the combination of WtD and CWD, providing a class of distance functions $WtCWD$ has shown a significant reduction in errors of approximation. The analysis of hyperspheres of $WtCWD$ provides more useful insights in other known classes of distance functions such as *m-neighbor distances*, *hyperoctagonal distances*, and *weighted t-cost distances*. This helps us in deriving theoretical expressions of MREs including the values with optimal scale adjustment. In our last and concluding chapter, we provide overall comparisons of different good digital distances in approximating Euclidean norms, and also discuss some of the open problems.

Chapter 6
Conclusion

In previous chapters, we review different classes of distance functions and discuss their properties in approximating Euclidean metrics. In several approaches for the convenience of mathematical treatment, the error analysis is carried out in a real space ($\mathscr{R}^n$) and mostly optimal results are obtained without considering integral nature of $\mathscr{L}^n$. Extending those results, we can search for good digital distances in the integral space. In this chapter, we conclude by summarizing these findings in the context of digital distances.

6.1 Choosing Good Digital Distances

One of the prime motivations in using digital distances is to compute distance transform, and then apply the distance transform in various applications of shape analysis [48, 49, 78], feature extraction [66], and feature representation [68]. A property that is looked for in such cases is the chamfering capability of such a distance function for computing the distance transform. This makes the computation efficient in $O(n)$ time complexity, where n is the number of grid points in the digital space. Not necessarily, every distance function has a chamfering mask. In a chamfering mask of a finite number of neighboring points around a point p, it needs to be ensured that all other remaining points in the space remain hidden or invisible from p, so that they cannot be reached by a path without passing through at least one of its neighbors. This is not satisfied when a distance function by itself has an infinite neighborhood. Even in the discrete space, with a finite number of neighbors within a given radius r, this may not be satisfied. For example, an Euclidean distance in a discrete grid may have finite number of points within a distance r, but given an infinite number of possible directions of forming a path from that point, it is impossible to design such a chamfering mask. This is the reason, we look for a digital distance which has the chamfering capability, as well as, is functionally very close to the Euclidean norm.

In previous chapters, we find that some of the distance functions such as $WtCWD_{isr\_eu}$, $WtCWD_{sub\_rec}$, WtD_{isr}, etc., have good approximation properties with Euclidean norms in any arbitrary dimension. But their chamfering capability is not yet known to us.[1] Out of the classes of distance functions having the property of chamfering, we observe that CWDs and HODs perform distinctly better than other classes such as m-neighbor[2] and t-cost[3] distances. Hence, we restrict our attention to finding good distances in classes such as CWD and HOD. Further, to keep the computation restricted to integer domain only, we consider integral weights of a CWD allowing a constant scale factor while approximating an Euclidean norm.

6.1.1   Choosing Integral CWDs

From Theorem 3.17, we get optimum set of weights for a CWD, which we identify as $CWD_{eu}^{(n)}$.[4] In 2-D, 3-D, and 4-D, close approximations of these weights are reported to be $< 3, 4 >$ [6], $< 3, 4, 5 >$ [6], and $< 3, 4, 5, 6 >$ [11], respectively (Sect. 3.5). In 4-D $< 7, 9, 11, 13 >$ is found to improve this approximation. Intuitively, we consider an integral weight assignment, which should closely follow proportional weights of $CWD_{eu}^{(n)}$. The assignment should also be such that it satisfies the conditions of Theorem 2.2 (or Theorem 2.3). Intuitively, we may generalize the $< 3, 4 >$, $< 3, 4, 5 >$, and $< 3, 4, 5, 6 >$ sequences to $< 3, 4, 5, \ldots, n + 2 >$ for a weight assignment to a CWD in n-D, and study the values of MREs from Theorem 4.33, at varying dimensions. It is trivially observed that these weight assignments satisfy the property of a norm, as the weights of equivalent LWDs are nonincreasing. Let us denote a distance function with such weight assignment in n-D, $CWD_{3\_4\_gen}^{(n)}$. However, as it has already been noted that in 4-D, there exists another distance function with better approximations of weights of $CWD_{eu}^{(n)}$. We consider another candidate set with the following weight assignment policy in the representational form of a LWD in n-D:

$$\gamma_i = \begin{cases} \lceil \frac{1}{\sqrt{n}-\sqrt{n-1}} \rceil & i = 1, \\ \lceil \gamma_1(\sqrt{i} - \sqrt{i-1}) \rceil & 2 \leq i \leq n. \end{cases} \tag{6.1}$$

In the above, γ_n is always kept at 1. As the weights of the above LWD is nonincreasing, they satisfy the property of the norm. We can convert this representation from LWD to CWD using Eq. (2.5). We refer to this distance function as $CWD_{eu\_int}^{(n)}$. In Fig. 6.1, plots of scale adjusted MRE for $CWD_{3\_4\_gen}^{(n)}$, $CWD_{eu\_int}^{(n)}$ and $CWD_{euopt}^{(n)}$ against the dimensions varying from 2 to 100 are shown. Though in the first few dimensional

---

[1]Whether they are chamferable or not, that remains an open question during the writing of this book.

[2]An m-neighbor distance is also a special case of a HOD.

[3]An t-cost distance is also a special case of a CWD.

[4]We ignore scale factor for proportional weight assignment in this case.

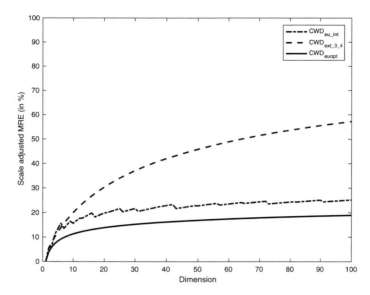

Fig. 6.1 Scale adjusted optimum MRE for $CWD^{(n)}_{3\_4\_gen}$, $CWD^{(n)}_{eu\_int}$ and $CWD^{(n)}_{euopt}$

spaces, $CWD^{(n)}_{3\_4\_gen}$ has a better performance than $CWD^{(n)}_{eu\_int}$, the latter has significantly improved MREs in dimension higher than 6. The trend of the variation of $CWD^{(n)}_{eu\_int}$ mimics that of optimal behavior of $CWD^{(n)}_{euopt}$. In Table 6.1, we show the values of scale adjusted MREs in dimensions from 2 to 10 for $CWD^{(n)}_{3\_4\_gen}$ and $CWD^{(n)}_{eu\_int}$. The optimal values of MRE as obtained from $CWD^{(n)}_{euopt}$ are also shown. As noted in the plots (Fig. 6.1), in low dimensional spaces $CWD^{(n)}_{3\_4\_gen}$ has lower MRE values than those of $CWD^{(n)}_{eu\_int}$, though in 3-D the latter has much improved performance. We note also in 2-D the integral CWDs of these two types are the same ($< 3, 4 >$). The lower MRE values among these two types of distances are highlighted by bold font. Naturally the optimum MREs obtained by $CWD^{(n)}_{euopt}$ are the lowest among these three distances in every dimension.

In Fig. 6.1, we show scale adjusted MREs for these two types of integral CWDs. In the figure, optimum MREs among distances in the class of CWD are also plotted by providing those values of $CWD^{(n)}_{euopt}$.

6.1.2 Choice of Hyperoctagonal Distances

In previous chapters, we discuss a few optimal results on error analysis for HODs in 2-D and 3-D. In Section 3.6, the results are available for simple octagonal distances in 2-D, and a few distances such as {1, 1, 2}, {1, 1, 2, 1, 2}, are found to have good approximation properties. In Sect. 3.7.2, using empirical error analysis, the neighbor-

Table 6.1 Scale adjusted optimum MRE for $CWD^{(n)}_{3\_4\_gen}$, $CWD^{(n)}_{eu\_int}$ and $CWD^{(n)}_{euopt}$ in n-D $(2 \le n \le 10)$

Dimension	$CWD^{(n)}_{3\_4\_gen}$			$CWD^{(n)}_{eu\_int}$			$CWD^{(n)}_{euopt}$
n	Distance	κ_{opt}	MRE_{sc}	Distance	κ_{opt}	MRE_{sc}	MRE_{opt}
2	$< 3, 4 >$	0.334	0.056	$< 3, 4 >$	0.334	0.056	0.040
3	$< 3, 4, 5 >$	0.326	0.079	$< 4, 6, 7 >$	0.233	**0.068**	0.060
4	$< 3, \ldots, 6 >$	0.318	**0.101**	$< 4, 6, 8, 9 >$	0.222	0.111	0.074
5	$< 3, \ldots, 7 >$	0.311	**0.121**	$< 5, 8, 10, 12, 13$	0.173	0.135	0.084
6	$< 3, \ldots, 8 >$	0.304	**0.139**	$< 5, 8, 10, 12, 14, 15 >$	0.169	0.157	0.092
7	$< 3, \ldots, 9 >$	0.298	0.156	$< 6, 9, 11, 13, 15, 17, 18 >$	0.144	**0.135**	0.098
8	$< 3, \ldots, 10 >$	0.293	0.172	$< 6, 9, 11, 13, 15, 17, 19, 20 >$	0.142	**0.150**	0.104
9	$< 3, \ldots, 11 >$	0.288	0.186	$< 6, 9, 11, 13, 15, 17, 19, 21, 22 >$	0.139	**0.165**	0.109
10	$< 3, \ldots, 12 >$	0.283	0.200	$< 7, 10, 13, 15, 17, 19, 21, 23, 25, 26 >$	0.121	**0.156**	0.113

hood sequences such as $\{1, 1, 2\}$ and $\{1, 2\}$ in 2-D, and $\{1, 1, 3\}$, $\{1, 1, 1, 1, 2, 2, 3\}$, and $\{1, 1, 1, 2, 3\}$ in 3-D are also found to be good approximating distances. In Sect. 5.6.2, the analysis is further extended on optimality conditions of MRE in 2-D and 3-D for HODs and those distances are reported to be close to the near-optimal points in solution spaces. The geometric approaches, discussed in Sect. 4.2.2, also corroborate these findings. But there are no such generalized results on optimal errors in any arbitrary dimension. That is why in Sect. 5.6.2.3, an empirical approach has been adopted to check all combinations of sorted neighborhood sequences of length at most L. A few such optimal neighborhood sequences till dimension 10 are listed in Table 5.5. By observing these trends, we consider a few general series of sorted sequences as candidates for studying MRE values with the expectation of having their good approximating properties. We chose these representations for sequence length 3, 4, and 5, and they are given by $[1^2 n^1]$, $[1^2 \lfloor \sqrt{n} \rfloor^1 n^1]$, and $[1^3 \lfloor \sqrt{n} \rfloor^1 n^1]$.[5] Our choices are made from the observations that we make in Table 5.5. In Fig. 6.2, plots of their scale adjusted MRE values are shown. We observe that for a high dimensional space, $[1^3 \lfloor \sqrt{n} \rfloor^1 n^1]$ provides lower MRE than other two candidate functions of a neighborhood sequence. In Table 6.2, the values of scale adjusted MRE along with the optimum scale for the distances are provided. We observe that in 2-D, and 3-D $\{1, 1, 2\}$, and $\{1, 1, 3\}$, respectively, have the least MRE values among their candidate peers. For other dimensions $(4 \le n \le 10)$, the distance functions providing the least error are also the same reported in Table 5.5. It vindicates our intuitive choice of functional descriptions of a neighborhood sequence in this context.

---

[5]Sometimes $[1^3 \lceil \sqrt{n} \rceil^1 n^1]$ are found to provide marginally better performance.

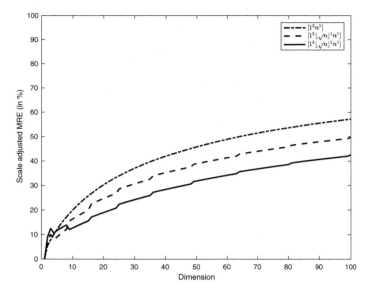

Fig. 6.2 Scale adjusted optimum MRE for $[1^2 n^1]$, $[1^2 \lfloor \sqrt{n} \rfloor^1 n^1]$, and $[1^3 \lfloor \sqrt{n} \rfloor^1 n^1]$

Table 6.2 Scale adjusted optimum MRE for $[1^2 n^1]$, $[1^2 \lfloor \sqrt{n} \rfloor^1 n^1]$, and $[1^3 \lfloor \sqrt{n} \rfloor^1 n^1]$ in n-D ($2 \leq n \leq 10$)

Dimension	$[1^2 n^1]$		$[1^2 \lfloor \sqrt{n} \rfloor^1 n^1]$		$[1^3 \lfloor \sqrt{n} \rfloor^1 n^1]$	
n	κ_{opt}	MRE_{sc}	κ_{opt}	MRE_{sc}	κ_{opt}	MRE_{sc}
2	0.9953	**0.0557**	0.9517	0.0767	0.9263	0.0917
3	1.0177	**0.0794**	0.9535	0.1010	0.9093	0.1250
4	1.0381	0.1010	1.0794	**0.0794**	0.9955	0.1061
5	1.0567	0.1208	1.0905	**0.0905**	0.9990	0.1170
6	1.0739	0.1390	1.1010	**0.1010**	1.0058	0.1245
7	1.0898	0.1559	1.1111	**0.1111**	1.0124	0.1319
8	1.1046	0.1716	1.1208	**0.1208**	1.0188	0.1390
9	1.1184	0.1862	1.1559	0.1559	1.0981	**0.1208**
10	1.1314	0.2000	1.1639	0.1639	1.1042	**0.1270**

6.2 Summary of Good Digital Distances Approximating Euclidean Norms

Finally, let us summarize a list of digital distances in a few low dimensional spaces, which have low ($\leq 5\%$) to moderate ($\leq 20\%$) MREs with respect to Euclidean norms. If results related to geometric errors are available, we also provide average π-error (average of volume-π, surface-π, and shape-π errors). In Table 6.3, we show these errors for some of the good digital distances in 2-D, 3-D, and 4-D reported by

Table 6.3   Good digital distances approximating Euclidean norms

n	Distance	κ_{opt}	MRE_{sc}	Avg-π error
2	$< 2, 3 >$	0.472	0.056	0.119
	$< 3, 4 >$	0.334	0.056	0.120
	$< 5, 7 >$	0.194	0.042	0.103
	$< 12, 17 >$	0.080	**0.040**	**0.098**
	$\{1, 2\}$	1.056	0.056	0.858
	$\{1, 1, 2\}$	0.995	0.056	0.099
	$\{1, 1, 1, 2\}$	0.952	0.078	0.178
	$\{1, 1, 1, 2, 2\}$	1.032	0.042	0.155
	$\mathscr{M}_{55}(4, 6, 9)$	0.243	0.029	
	$\mathscr{M}_{55}(5, 7, 11)$	0.200	0.018	
	$\mathscr{M}_{55}(9, 13, 20)$	0.110	0.015	
	$\mathscr{M}_{55}(17, 24, 38)$	0.058	0.014	
3	$< 2, 3, 4 >$	0.450	0.101	0.349
	$< 3, 4, 5 >$	0.326	0.079	0.261
	$< 4, 6, 7 >$	0.233	0.068	0.247
	$< 7, 10, 12 >$	0.135	0.064	0.213
	$< 19, 27, 33 >$	0.049	**0.061**	**0.175**
	$\{1, 1, 3\}$	1.018	0.079	0.232
	$\{1, 1, 2, 3\}$	1.068	0.068	0.614
	$\{1, 1, 1, 2, 3\}$	1.002	0.085	0.288
	$\{1, 1, 1, 1, 2, 2, 3\}$	0.993	0.095	0.598
	$\mathscr{M}_{555}(4, 6, 7, 9, 10, 12)$	0.243	0.029	
	$\mathscr{M}_{555}(9, 13, 16, 20, 22, 27)$	0.109	0.027	
	$\mathscr{M}_{555}(20, 29, 35, 45, 49, 60)$	0.049	0.024	
4	$< 3, 4, 5, 6 >$	0.318	0.101	0.226
	$< 4, 6, 8, 9 >$	0.222	0.111	0.291
	$< 7, 9, 11, 13 >$	0.141	0.103	0.214
	$\{1, 3\}$	1.101	0.101	
	$\{1, 1, 4\}$	1.038	0.101	
	$\{1, 1, 2, 4\}$	1.079	**0.079**	
	$\{1, 1, 1, 2, 4\}$	0.996	0.1061	

researchers. In addition, we also provide them for a few more dimensions (5 to 10) in Table 6.4, from the study presented in this book. It is expected that larger the mask size of a chamfering norm and closer the weights to Euclidean lengths of the vectors in the mask, better would be the approximation of the Euclidean norm. This is also corroborated by the results reported in 2-D and 3-D [12, 76, 77, 80] . But it is difficult to derive a functional form of these distances. There CWDs and HODs have an advantage. This is the reason we kept CMIDs in Table 6.3, out of comparison of performances from those of other two classes, namely CWD and HOD. We highlight

Table 6.4 Good digital distances approximating Euclidean norms

n	Distance	κ_{opt}	MRE_{sc}	Avg.-π error
5	$< 3, 4, 5, 6, 7 >$	0.311	0.121	0.354
	$< 5, 8, 10, 12, 13 >$	0.173	0.135	0.471
	$\{1, 1, 5\}$	1.057	0.121	
	$\{1, 1, 2, 5\}$	1.091	**0.090**	
	$\{1, 1, 1, 2, 5\}$	0.999	0.117	
	$\{1, 1, 1, 3, 5\}$	1.074	0.092	
6	$< 3, 4, 5, 6, 7, 8 >$	0.304	0.139	0.329
	$< 5, 8, 10, 12, 14, 15 >$	0.169	0.157	0.443
	$\{1, 1, 6\}$	1.074	0.139	
	$\{1, 1, 2, 6\}$	1.101	**0.101**	
	$\{1, 1, 1, 2, 6\}$	1.006	0.130	
	$\{1, 1, 1, 3, 6\}$	1.079	**0.101**	
7	$< 3, 4, 5, 6, 7, 8, 9 >$	0.298	0.156	0.442
	$< 6, 9, 11, 13, 15, 17, 18 >$	0.114	0.135	0.434
	$\{1, 1, 7\}$	1.089	0.156	
	$\{1, 1, 2, 7\}$	1.111	0.111	
	$\{1, 1, 1, 2, 7\}$	1.0124	0.132	
	$\{1, 1, 1, 3, 7\}$	1.085	**0.108**	
8	$< 3, 4, 5, 6, 7, 8, 9, 10 >$	0.293	0.172	0.421
	$< 6, 9, 11, 13, 15, 17, 19, 20 >$	0.142	0.150	0.413
	$\{1, 1, 8\}$	1.105	0.172	
	$\{1, 1, 2, 8\}$	1.121	0.121	
	$\{1, 1, 1, 2, 8\}$	1.019	0.139	
	$\{1, 1, 1, 3, 8\}$	1.092	**0.114**	
9	$< 3, 4, 5, 6, 7, 8, 9, 10, 11 >$	0.288	0.186	0.523
	$< 6, 9, 11, 13, 15, 17, 19, 21, 22 >$	0.139	0.165	0.505
	$\{1, 1, 9\}$	1.118	0.186	
	$\{1, 1, 3, 9\}$	1.156	0.156	
	$\{1, 1, 1, 3, 9\}$	1.098	**0.121**	
10	$< 3, 4, 5, 6, 7, 8, 9, 10, 11, 12 >$	0.283	0.200	0.500
	$< 7, 10, 13, 15, 17, 19, 21, 23, 25, 26 >$	0.120	0.156	0.428
	$\{1, 1, 10\}$	1.131	0.200	
	$\{1, 1, 3, 10\}$	1.164	0.164	
	$\{1, 1, 1, 3, 10\}$	1.042	**0.127**	
	$\{1, 1, 1, 4, 10\}$	1.128	0.140	

the least errors among the distances of these two classes in the tables. As the *average* π *-error* is computed for both these types of distances functions in 2-D and 3-D, we also highlight the least error among them. In higher dimensional spaces, we still do not have related results for HODs, and we refrain from making any such comparison based on geometric errors.

From these observations, we note that in 2-D, the distance $< 12, 17 >$ is a very good candidate for approximating Euclidean norm. It has the least MRE and the average-π error within the distances belonging to the classes, CWD and HOD, considered in Table 6.3. In 3-D also, a similar feature is observed for the distance $< 19, 27, 33 >$. But in all other higher dimensional spaces, distances from the hyperoctagonal class have better performances than those belonging to the class CWD.

6.3   A Few Open Questions

From this study, a few open questions may be noted before we conclude this discussion.

1. The chamfering capabilities of a $WtCWD$ may need to be explored further. As we can compute the vertices of a hypersphere of a distance function of this class (Theorem 5.4), it is possible to get a set of vectorial directions for chamfering. Relating them to integral points with associated weights for deriving the generator of a chamfer mask (Definition 2.6) may be worth trying. It may be necessary to look for the properties which could characterize whether the hypersphere of a $WtCWD$ can form an *equivalent rational ball* (ERB) (Definition 2.9).

2. We could derive theoretical expressions for volume and surface of the hypersphere of a CWD in any arbitrary dimension (Theorems 4.13 and 4.14). We have a few related results in 2-D and 3-D for special type of WtDs (Sect. 4.4.5) and HODs (Sect. 4.2.2.1). This approach may also be extended for CMIDs in these two lower dimensional spaces. But we are yet to get any result in higher dimensions for all these distances.

3. There are a few analyses to derive closed form expressions of CMIDs, and relating the vectors and weights in a mask to a norm in 2-D (Sect. 2.3.1.1), and 3-D (Sect. 2.3.1.2). They need to be extended to general cases. Similarly, the error analysis for any arbitrary CMID norm may be developed (Sect. 3.5.5).

Besides above theoretical questions, there still exists the critical concern of a pragmatist to judge the usefulness of these distance functions, and also the utility of exercises that follow to characterize their proximity to an Euclidean norm.

6.4 Concluding Remarks

Finally, we come to the end of this study. After going through all these mathematical notations, theorems, lemmas, equations, plots, figures, and tables in different chapters, let us take a pause, and pretend to be a philosopher with wild dreams and conjectures without any shouting and swearing! Our understanding of this world or universe, whatever you may say, is an abstraction from a mathematical model. We know that this apparent Euclidean world has a different geometric model over time and space. Whether a discrete and digital model has a role there, neither we can pledge, nor we can overrule!

References

1. Arcelli, C., & Sanniti Di Baja, G. (1988). Finding local maxima in a pseudo-Euclidean distance transform. *Computer Vision, Graphics and Image Processing, 43*, 361–367.
2. Barni, M., Buti, F., Bartolini, F., & Cappellini, V. (2000). A quasi-Euclidean norm to speed up vector median filtering. *IEEE Transaction on Image Processing, 9*(10), 1704–1709. October.
3. Barni, M., Cappellini, V., & Mecocci, A. (1994). Fast vector median filter based on Euclidean norm approximation. *IEEE Signal Processing Letters, 1*(6), 92–94. June.
4. Berger, M. (1978). Convexes et polytopes, polyedres reguliers, et volumes (vol. 3, Cedic/Fernand Nathan, 2 eme ed.) 11.8.12.
5. Blum, H. (1964, November). In Dunn, W. (Ed.), *A transformation for extracting new descriptors of shape. Perception of speech and visual form* (pp. 362–380).
6. Borgefors, G. (1984). Distance transformations in arbitrary dimensions. *Computer Vision, Graphics and Image Processing, 27*, 321–345.
7. Borgefors, G. (1986). Distance transformations in digital images. *Computer Vision, Graphics and Image Processing, 34*, 344–371.
8. Borgefors, G. (1991). Another comment on "a note on distance transformations in digital images ". *Computer Vision, Graphics and Image Processing: Image Understanding, 54*(2), 301–306.
9. Borgefors, G. (1993). Centres of maximal discs in the 5-7-11 distance transforms. In: *Proceedings of the 8th Scandinavian Conference* (pp. 105–111).
10. Borgefors, G. (1996). On digital transforms in three dimensions. *Computer Vision and Image Understanding, 64*, 368–376.
11. Borgefors, G. (2003). Weighted digital distance transforms in four dimensions. *Discrete Applied Mathematics, 125*, 161–176.
12. Butt, M. A., & Maragos, P. (1998). Optimum design of chamfer distance transforms. *IEEE Transaction on Image Processing, 7*(10), 1477–1483.
13. Celebi, M. E., Celiker, F., & Kingravi, H. A. (2011). On Euclidean norm approximations. *Pattern Recognition, 44*, 278–283.
14. Celebi, M. E., Kingravi, H. A., & Celiker, F. (2012). Comments on "on approximating Euclidean metrics by weighted t-cost distances in arbitrary dimension". *Pattern Recognition Letters, 33*, 1422–1425.
15. Chaudhuri, D., Murthy, C. A., & Chaudhuri, B. B. (1992). A modified metric to compute distance. *Pattern Recognition, 25*(7), 667–677.
16. Coeurjolly, D., & Montanvert, A. (2007). Optimal separable algorithms to compute the reverse Euclidean distance transformation and discrete medial axis in arbitrary dimension. *IEEE Transaction on Pattern Analysis and Machine Intelligence, 29*(3), 1–12.
17. Coeurjolly, D., & Sivignon, I. (2020). Efficient distance transformation for path-based metrics. *Computer Vision and Image Understanding, 194*(2020), 102925.

© The Author(s), under exclusive license to Springer Nature Singapore Pte Ltd. 2020
J. Mukhopadhyay, *Approximation of Euclidean Metric by Digital Distances*,
https://doi.org/10.1007/978-981-15-9901-9

18. Das, P. P. (1988). *Paths and distances in digital geometry*. Ph.D. dissertation, Indian Institute of Technology, Kharagpur, India.
19. Das, P. P. (1992, November 15–20). Hypersphere of n-sequence distances. In *Proceedings of SPIE Conference on Vision Geometry*, Boston, USA (vol. 1832, pp. 61–67)
20. Das, P. P. (1992). Best simple octagonal distances in digital geometry. *Journal of Approximation Theory, 68*(2), 155–174. February.
21. Das, P.P. (1992, November 15–20). The real m-neighbor distance. In *Proceedings of SPIE Conference on Vision Geometry*, Boston, USA (vol. 1832, pp. 68–78).
22. Das, P. P. (1992). A note on "distance functions in digital geometry". *Information Sciences, 64*, 181–190.
23. Das, P. P., Chakrabarti, P. P., & Chatterji, B. (1987). Distance functions in digital geometry. *Information Sciences, 42*, 113–136.
24. Das, P. P., Chakrabarti, P. P., & Chatterji, B. (1987). Generalized distances in digital geometry. *Information Sciences, 42*, 51–67.
25. Das, P. P., Chakrabarti, P. P., & Chatterji, B. (1991). The t-cost-m-neighbour distance in digital geometry. *Journal of Geometry, 42*, 42–58.
26. Das, P. P., & Chatterji, B. N. (1988). Knight's distance in digital geometry. *Pattern Recognition Letters, 7*, 215–226.
27. Das, P. P., & Chatterji, B. N. (1989). Estimation of errors between Euclidean and m-Neighbor distance. *Information Sciences, 48*, 1–26.
28. Das, P. P., & Chatterji, B. N. (1990). Octagonal distances for digital pictures. *Information Sciences, 50*, 123–150.
29. Das, P. P., & Chatterji, B. N. (1990). Hyperspheres in digital geometry. *Information Sciences, 50*, 73–93.
30. Das, P. P., & Mukherjee, J. (1990). Metricity of super-knight's distance in digital geometry. *Pattern Recognition Letters, 11*, 601–604.
31. Das, P. P., Mukherjee, J., & Chatterji, B. N. (1992). The t-cost distance in digital geometry. *Information Sciences, 59*, 1–20.
32. Danielsson, P. E. (1993). 3-D octagonal metrics. In: *Proceedings of Eighth Scandinavian Conference* (pp. 727–736).
33. Fabbri, R., da Fontour Costa, L., Torelli, J. C., & Bruno, O. M. (2008). 2D Euclidean distance transform algorithms: A comparative survey. *ACM Computing Survey, 40*, 1–44.
34. Farkas, J., Bajak, S., & Nagy, B. (2006). Notes on approximating the Euclidean circle in square grids. *Pure Mathematics and Applications, 17*, 309–322.
35. Fouard, C., Strand, R., & Borgefors, G. (2007). Weighted distance transforms generalized to modules and their computation on point lattices. *Pattern Recognition, 40*(9), 2453–2474.
36. Hajdu, A., & Hajdu, L. (2004). Approximating the Euclidean distance using non-periodic neighborhood sequences. *Discrete Applied Mathematics, 283*, 101–111.
37. Jones, M. W., Baerentzen, J. A., & Sramek, M. (2006). 3D Distance fields: A survey of techniques and applications. *IEEE Transaction on Visualization and Computer Graphics, 12*, 581–599.
38. Kumar, M. A., Mukherjee, J., Chatterji, B. N., & Das, P. P. (1995) A geometric approach to obtain best octagonal distances. In *9th Scandinavian Conference on Image Processing* (pp. 491–498).
39. Kumar, M. A., Mukherjee, J., Chatterji, B. N., & Das, P. P. (1996). Representation of 2-D and 3-D binary images using medial circles and spheres. Intl. *Journal of Pattern Recognition and Artificial Intelligence, 10*, 365–387.
40. Lam, L., Lee, S. W., & Suen, C. Y. (1992). Thinning methodologies: A comprehensive survey. *IEEE Transaction on Pattern Recognition and Machine Intelligence, 14*(9), 869–885.
41. Montanari, U. (1968). A method for obtaining skeletons using a quasi-Euclidean distance. *Journal of Association of Computing Machinery, 15*(4), 600–624.
42. Mukherjee, J. (2011). On approximating Euclidean metrics by weighted t-cost distances in arbitrary dimension. *Pattern Recognition Letters, 32*, 824–831.

43. Mukherjee, J. (2013). Hyperspheres of weighted distances in arbitrary dimension. *Pattern Recognition Letters, 34*, 117–123.
44. Mukherjee, J. (2013). Linear combination of norms in improving approximation of Euclidean norm. *Pattern Recognition Letters, 34*, 1348–1355.
45. Mukherjee, J. (2014). Linear combination of weighted t-cost and chamfering weighted distances. *Pattern Recognition Letters, 40*(2014), 72–79.
46. Mukherjee, J. (2016). Error analysis of octagonal distances defined by periodic neighborhood sequences for approximating Euclidean metrics in arbitrary dimension. *Pattern Recognition Letters, 75*(2016), 16–23.
47. Mukherjee, J., Das, P. P., Kumar, M. A., & Chatterji, B. N. (2000). On approximating Euclidean metrics by digital distances in 2-D and 3-D. *Pattern Recognition Letters, 21*, 573–582.
48. Mukherjee, J., Kumar, M. A., Das, P. P., & Chatterji, B. N. (2000). Fast computation of cross-sections of 3-D objects from their medial axis transforms. *Pattern Recognition Letters, 21*, 605–613.
49. Mukherjee, J., Kumar, M. A., Das, P. P., & Chatterji, B. N. (2002). Use of medial axis transforms for computing normals at boundary points. *Pattern Recognition Letters, 23*, 1649–1656.
50. Mukhopadhyay, J., Das, P. P., Chattopadhyay, S., Bhowmick, P., & Chatterji, B. N. (2013). *Digital geometry in image processing.* CRC Press.
51. Muller, M. E. (1959). A note on a method for generating points uniformly on N-dimensional spheres. *Communications of the ACM, 2*(4), 19–20.
52. Nagy, B. (2003). Distance functions based on neighborhood sequences. *Publicationes Mathematicae Debrecen, 63*(3), 483–493.
53. Nagy, B. (2008). Distance with generalized neighborhood sequences in n-D and ∞-D. *Discrete Applied Mathematics, 156*, 2344–2351.
54. Nagy, B., & Strand, R. (2011). Approximating Euclidean circles by neighborhood sequences in a hexagonal grid. *Theoretical Computer Science, 412*, 1364–1377.
55. Nagy, B., Strand, R., Normand, N., 2013. A weight sequence distance function. In *11th International Symposium on Mathematical Morphology (ISMM 2013)*, Uppsala, Sweden. LNCS. (vol. 7883, pp. 292–301).
56. Nicolas Normand, N., & Evenou, P. (2009). Medial axis look up table and test neighborhood computation for 3-D chamfer norms. *Pattern Recognition, 42*(2009), 2288–2296.
57. Normand, N., Strand, R., Evenou, P., & Arlicot, A. (2011). Path-based distance with varying weights and neighborhood sequences. In *16th IAPR International Conference on Discrete Geometry for Computer Imagery (DGCI 2011)*, Nancy, France, LNCS (vol. 6607, pp. 199–210).
58. Ohashi, Y. (1994). Fast linear approximations of Euclidean distance in higher dimensions. In P. Heckbert (Ed.), *Graphics gems IV.* Academic.
59. Okabe, N., Toriwaki, J., & Fukumura, T. (1983). Paths and distance functions on three-dimensional digitized pictures. *Pattern Recognition Letters, 1*, 205–212.
60. Ragnemalm, I. (1993). The Euclidean distance transform in arbitrary dimensions. *Pattern Recognition Letters, 14*(1993), 883–888.
61. Remy, E., & Thiel, E. (2000). July). Optimizing 3-D chamfer masks with norm constraints. In R. Malgouyres (Ed.), *7th IWCIA, International Workshop on Combinatorial Image Analysis* (pp. 39–56). Caen:
62. Rhodes, F. (1995). On the metrics of Chaudhuri, Murthy and Chaudhuri. *Pattern Recognition, 28*(5), 745–752.
63. Rosenfeld, A. (1981). Three-dimensional digital topology. *Information and Control, 50*, 119–127.
64. Rosenfeld, A. A., & Pfaltz, J. L. (1966). Sequential operations in digital picture processing. *Journal of Association of Computing Machinery, 4*, 471–494.
65. Rosenfeld, A. A., & Pfaltz, J. L. (1968). Distance functions in digital pictures. *Pattern Recognition, 1*, 33–61.
66. Sanniti di Baja, G., & Svensson, S. (2002). A new shape descriptor for surfaces in 3-D images. *Pattern Recognition Letters, 23*, 703–711.

67. Seol, C., & Cheun, K. (2008). A low complexity Euclidean norm approximation. *IEEE Transaction on Signal Processing, 56*(4), 1721–1726.
68. Strand, R. (2011). Sparse object representations by digital distance functions. In Debled-Rennesson et al. (Eds.), *DGCI 2011*. LNCS (vol. 6607, pp. 211–222).
69. Strand, R. (2007). Weighted distances based on neighborhood sequences. *Pattern Recognition Letters, 28*, 2029–2036.
70. Strand, R. (2007). Weighted distances based on neighborhood sequences for point-lattices. *Discrete Applied Mathematics, 157*, 641–652.
71. Strand, R., & Borgefors, G. (2005). Distance transforms for three-dimensional grids with non-cubic voxels. *Computer Vision and Image Understanding, 100*, 294–311.
72. Strand, R., & Nagy, B. (2006). Approximating Euclidean distance using distances based on neighborhood sequences in non-standard three-dimensional grids. In *IWCIA 2006*. LNCS (vol. 4040, p. 89).
73. Strand, R., & Nagy, B. (2007). Distances based on neighborhood sequences in non-standard three-dimensional grids. *Discrete Applied Mathematics, 155*, 548–557.
74. Strand, R., & Nagy, B. (2008). Weighted neighborhood sequences in non-standard three-dimensional grids-parameter optimization. In *IWCIA 2008*. LNCS (vol. 4958, pp. 51–62).
75. Strand, R., Nagy, B., Fouard, C., & Borgefors, G. (2006). Generating distance maps with neighborhood sequences. In *DGCI 2006*. LNCS (vol. 4245, pp. 295–307).
76. Svensson, S., & Borgefors, G. (2002). Digital distance transforms in 3-D images using information from neighborhoods up to $5 \times 5 \times 5$. *Computer Vision and Image Understanding, 88*, 24–53.
77. Svensson, S., & Borgefors, G. (2002). Distance transforms in 3-D using four different weights. *Pattern Recognition Letters, 23*(2002), 1407–1418.
78. Svenssona, S., & Sanniti di Baja, G. (2002). Using distance transforms to decompose 3-D discrete objects. *Image and Vision Computing, 20*(2002), 529–540.
79. Thiel, E. (2001, December 2). Geometrie des distances dechanfrein, memoire d'habilitation a diriger des recherches, Universite de la Mediterranee, Aix-Marseille. http://www.lif-sud.univ-mrs.fr/~thiel/hdr/.
80. Verwer, B. J. H. (1991). Local distances for distance transformations in two and three dimensions. *Pattern Recognition Letters, 12*, 671–682.
81. Wang, J., & Tan, Y. (2013). Efficient Euclidean distance transform algorithm of binary images in arbitrary dimensions. *Pattern Recognition, 46*, 230–242.
82. Yamashita, M., & Honda, N. (1984). Distance functions defined by variable neighborhood sequences. *Pattern Recognition, 17*(5), 509–513.
83. Yamashita, M., & Ibaraki, T. (1986). Distances defined by neighborhood sequences. *Pattern Recognition, 19*(3), 237–246.
84. Yokoi, S., Toriwaki, J., & Fukumara, T. (1981). On generalized distance transformation of digitized pictures. *IEEE Transaction on Pattern Analysis and Machine Intelligence, 3*(4), 424–443.

Index

Printed in the United States
By Bookmasters